Arsenic Removal Technologies

Akhilesh Kumar Yadav · Saba Shirin ·
Monika Simon · Ankit Kumar · Anil Kumar Dikshit
Editors

Arsenic Removal Technologies

From Challenges to Innovative Solutions

Springer

Editors
See next page

ISBN 978-3-031-91170-5 ISBN 978-3-031-91171-2 (eBook)
https://doi.org/10.1007/978-3-031-91171-2

This Springer imprint is published by the registered company Springer Nature Switzerland AG
The registered company address is: Gewerbestrasse 11, 6330 Cham, Switzerland

Editors
Akhilesh Kumar Yadav
Department of Environmental Engineering and Management
Chaoyang University of Technology
Taichung, Taiwan

Department of Electronics and Communication Engineering
Manipal University Jaipur
Jaipur, India

Department of Civil Engineering
Manipal University Jaipur
Jaipur, India

Department of Mining Engineering
Indian Institute of Technology (Banaras Hindu University)
Varanasi, India

Environmental Science and Engineering Department
Indian Institute of Technology Bombay
Mumbai, India

Monika Simon
Department of Hydrology
Indian Institute of Technology Roorkee
Roorkee, India

Department of Civil Engineering
SRM University Andhra Pradesh
Mangalagiri, India

Anil Kumar Dikshit
Environmental Science and Engineering Department
Indian Institute of Technology Bombay
Mumbai, India

Saba Shirin
Department of Environmental Science
School of Vocational Studies and Applied Sciences
Gautam Buddha University
Greater Noida, India

Department of Mining Engineering
Indian Institute of Technology (Banaras Hindu University)
Varanasi, India

Ankit Kumar
Department of Chemical Engineering
Indian Institute of Technology Gandhinagar
Gandhinagar, India

Lohum Cleantech Pvt. Ltd.
Greater Noida, India

Dedicated
to
Environmental Engineers and Scientists
around the World

Foreword

Arsenic contamination in water resources remains one of the most pressing global environmental and public health challenges. Its presence in drinking water poses serious health risks, including chronic diseases, developmental disorders, and life-threatening conditions. Addressing this issue requires a multifaceted approach that combines scientific innovation, sustainable technologies, and regulatory frameworks to ensure safe and quality water accessible for all. This book is edited by Akhilesh Kumar Yadav, Saba Shirin, Monika Simon, Ankit Kumar, and Anil Kumar Dikshit who each have expertise in the subject area and have assembled a team of impressive contributors. It comprises 14 chapters providing a comprehensive exploration of arsenic occurrence, its health implications, and the latest advances in its removal through bioremediation and technological solutions. The chapters systematically cover fundamental aspects, including the impact of arsenic exposure on human health, existing regulations for safe drinking water, and conventional and emerging remediation techniques. Particular emphasis is placed on bioremediation technologies that utilize microorganisms and natural plant-based solutions for the removal of arsenic. Further, the book delves into the potential of low-cost adsorption materials, ion exchange, and coagulation techniques as viable alternatives for mitigating arsenic pollution.

The strengths and limitations of conventional methods are critically examined, paving the way for future research and technological advancements. By presenting state-of-the-art research and innovative approaches, this book will serve as a valuable resource for scientists, policymakers, engineers, and environmentalists dedicated to addressing arsenic contamination. It not only contributes to scientific discourse, but also provides actionable solutions that can lead to improved public health outcomes and environmental sustainability.

The editors and contributors are to complimented for their dedication to this critical issue and for compiling such an insightful and forward-thinking volume. It is my hope that this book will inspire further research, technological innovations, and collaborative efforts to combat arsenic pollution and protect water quality worldwide.

Prof. Vijay P. Singh
Ph.D., D.Sc., D.Eng. (HC, Waterloo), D.Sc. (HC, Guelph), Ph.D. (HC, Basalicata), D.Sc. (HC, McGill), P.E., P.H., BC WRE (Hon.), D.EWRA (Hon.), Dist.M. ASCE, Hon.Dist. M. IWRA, Dist.F. AGGS, Hon.M. AWRA, F.AGU, EASA, Academician (GFA), NAE
Distinguished Professor
Regents Professor
Caroline and William N. Lehrer Distinguished Chair in Water Engineering
Department of Biological and Agricultural Engineering
Zachry Department of Civil and Environmental Engineering
Texas A&M University
College Station, TX, USA

Preface

In recent years, there has been significant progress in the use of various sustainable nanoparticles to remediate contaminated water. Arsenic contamination, arising from both natural and manmade sources, poses a global threat to water resources. In developing countries, inadequate water treatment infrastructure intensifies the health risks associated with arsenic exposure through drinking water. Groundwater is relatively more vulnerable to arsenic contamination from geothermal input, including the weathering and dissolution of arsenic-bearing rocks and minerals. Arsenic mobility increases when there is no oxygen in the water, making groundwater more vulnerable to arsenic contamination. Arsenic contamination in groundwater poses a severe public health risk, with ingestion linked to cancer, cardiovascular diseases, neurological disorders, and other life-threatening conditions. Chronic exposure to arsenic concentrations beyond 10 μg/L can cause severe skin disorders, organ damage, and an increased risk of lung and bladder cancer. Addressing this crisis requires innovative and sustainable water treatment solutions that meet stringent quality standards. Therefore, we need advanced water treatment methods to meet stringent quality standards.

This book explores alternative methods of arsenic contamination, its health impacts, and advanced removal strategies. Beginning with an overview of arsenic occurrence and its adverse effects on human health, the chapters delve into diverse remediation techniques, including biological and plant-based bioremediation, nanotechnology-driven solutions, and low-cost adsorption materials. Traditional methods such as ion exchange, coagulation, and electrodialysis are also critically evaluated for their effectiveness and limitations. Furthermore, the book highlights the significance of sustainable and integrated bioremediation technologies, emphasizing their potential for large-scale implementation with minimal environmental impact. By bringing together the latest research and insights from leading experts, this book aims to guide scientists, engineers, policymakers, and environmentalists in developing practical and efficient arsenic remediation strategies. We hope that the knowledge shared within these pages will contribute to innovative solutions and inspire further research in the field of arsenic remediation.

We first like to thank the Almighty for His blessings throughout the completion of this project, enabling us to explore this challenging research domain with a technical perspective successfully. The authors are sincerely grateful to Springer for the opportunity to publish this work on such a contemporary topic of modelling and simulations. Special thanks go to all the contributors who made this volume a valuable source of knowledge, reporting the latest findings in their respective fields. Of the 36 abstracts submitted, 25 were selected for full-chapter contributions, and 14 chapters were ultimately included based on the scope and quality of the submissions. The authors have shown tremendous effort throughout the process, from writing and revising based on comments to addressing Springer's evaluation reports and final proofreading. The review process adhered to a double-blind method, in which both authors and reviewers' identities were kept confidential to ensure unbiased decision making. We also thank the Springer publication team for their timely responses and guidance.

The editors welcome any feedback, comments, or suggestions to improve future editions.

Taichung, Taiwan — Akhilesh Kumar Yadav, Ph.D.
Greater Noida, India — Saba Shirin, Ph.D.
Roorkee, India — Monika Simon, Ph.D.
Greater Noida, India — Ankit Kumar, Ph.D.
Mumbai, India — Anil Kumar Dikshit, Ph.D.

Contents

About the Editors

Akhilesh Kumar Yadav earned a Bachelor of Technology in Electronics and Communication Engineering from Chaudhary Charan Singh University in Meerut, India. He then obtained a Master of Technology in Environmental Engineering from Madan Mohan Malaviya Engineering College in Gorakhpur, India (affiliated with Dr. A. P. J. Abdul Kalam Technical University, Lucknow). His doctoral degree, with a research focus on air pollution, was awarded by the Indian Institute of Technology (Banaras Hindu University), Varanasi, India. Dr. Yadav has served at various institutions, including Indian Institute of Technology (Banaras Hindu University), Varanasi, India; Indian Institute of Technology Bombay, Mumbai, India; Bhabha Atomic Research Centre, Mumbai, India (as a research collaborator); Manipal University Jaipur, Jaipur, India and Chaoyang University of Technology in Taichung, Taiwan. He has authored/co-authored several research and review articles, book chapters, and edited/authored books, and holds a patent. Dr. Yadav received the Young Engineer Award from the Institution of Engineers (India) and the Young Scientist Award from VDGOOD Technology Factory, Kolkata. His research interests include air, water, and soil pollution; climate change; vulnerability; human health risk assessments; and GIS applications in environmental pollution and management. He is an active member of several reputed international professional bodies, including the IEI (Kolkata), ECI (New Delhi), AEACI (Mumbai), MSI (New Delhi), and IASTA (Mumbai).

Saba Shirin completed her doctoral degree at the Indian Institute of Technology (Banaras Hindu University) Varanasi, India, in 2019. Presently, she is working as an Assistant Professor (OCFD), at Gautam Buddha University, Greater Noida, India. Her research interests are water and wastewater treatment, air quality, risk assessment, environmental pollution, environmental monitoring, and solid waste management. She has published articles in peer-reviewed journals and was a reviewer in reputed journals. She has attended many international and national conferences and workshops around the world. She is an Associate Member of the Association of Environmental Analytical Chemistry of India, Analytical Chemistry Division, Bhabha Atomic Research Centre, Mumbai, India.

Monika Simon holds a Ph.D. in Environmental Science and Engineering from the Indian Institute of Technology Roorkee, India. She is currently serving as a Postdoctoral Researcher at SRM University-AP, Andhra Pradesh. Prior to this, she worked as a Research Associate on an international collaborative project at IIT Roorkee. She earned her B.Tech. in Biotechnology from Sam Higginbottom University of Agriculture, Technology, and Sciences, Allahabad, and her M.Tech. in Environmental Engineering from Madan Mohan Malaviya University of Technology, Gorakhpur. She also brings over four years of industrial experience in the field of wastewater treatment. Dr. Simon has authored several peer-reviewed research articles and book chapters and holds a patent related to innovative wastewater treatment solutions. Her research interests include microbial bioremediation, biodegradation of microplastics, and the development of sustainable and frugal technologies for water and wastewater treatment.

Ankit Kumar currently working as R&D Manager at Lohum Cleantech Private Limited. Here, I am working on efficient extraction of 9999 PPT purity of cobalt/ Nickel and Rare earth metals (REM) from industrial inorganic byproducts or sludge. He has completed his B.Tech. M.Tech. and Ph.D. in chemical Engineering. Completing his Ph.D. in Chemical Engineering at the Indian Institute of Technology (Banaras Hindu University). He has dedicated his research efforts to advancing the utilization of biomass for energy production. His work primarily focused on catalyst development, particularly Metal Organic Framework (MOF)-derived Ni-based catalysts, for efficient hydrogen (H_2) production from bio-oil and its model compounds in a packed bed reactor. He also worked as Post-Doctoral Research Fellow at IIT GN-India. Here, he worked in green catalysis synthesis and its activity test, specifically concentrating on catalysts for H_2 production from formic acid/methanol decomposition at room temperature. He carried out a systematic study that shows how the metal-support interface sites affect the catalytic performance of Pd catalysts in terms of activity and selectivity. Dr. Ankit has also published more than a dozen peer review articles, number of book chapters and patents during his research journey.

Anil Kumar Dikshit is Higher Administrative Grade (HAG) Professor in the Environmental Infrastructure and Clean Tech (EICT) Group and In-charge of GIS Laboratory for Environmental Management and Sustainability (GEMS) of the Environmental Science and Engineering (ESE) Department at Indian Institute of Technology, Bombay, India.

He did his graduation and masters' in Civil Engineering in 1985 and 1987, respectively and completed Ph.D. in Civil and Environmental Engineering from Cornell University, USA in the year 1994. The areas of his teaching interest are environmental systems planning and management; environmental impact assessment; water supply and wastewater engineering; environmental engineering; and solid and biomedical waste management while the areas of research interest include environmental infrastructure; clean technology; urban, biomedical, hazardous and electronic solid waste management; environmental modeling, optimization, impact assessment, sustainability and life cycle assessment; and GIS applications to water and environment related problems.

He has more than 30 years of experience in Teaching, Research and Consultancy at IIT Kanpur, Cornell University, IIT Kharagpur and IIT Bombay. He has also taken several short assignments as visiting professor at Kigali Institute of Science and Technology, Rwanda; University of Kwazulu-Natal, South Africa; Nanyang Technological University, and National University of Singapore, Singapore; Malardalen University, Sweden; and Asian Institute of Technology, Bangkok. He has guided 40 masters' and 22 doctoral scholars while six research scholars are working on their Ph.Ds. He has also worked on more than 150 national and international research and consultancy projects. He is having research collaboration with German, French, USA, Singapore, South Africa, Sweden, Norway and Thailand Universities and Institutions and has organized 10 International Conferences, eight Workshops and 30 Training Courses. He is prolific speaker and has delivered more than 300 invited talks at various national and international forums.

The results of academic research work have provided one patent, eight books, 11 chapters in edited books, 100 reports, 150 publications in international and national journals, 140 in international and national conferences while many more papers are under review. In addition, more than 20 prototypes/equipment related to various environmental pollution management topics; academic versions of more than 10 environmental modeling and simulation software(s); and about 7 GIS applications on environmental management problems have been developed under his guidance. He is associate editor, reviewer or editorial board member for many peer-reviewed environmental journals.

Chapter 1
Overview on Occurrence of Arsenic on Human Health and Its Removal Through Microorganisms

Louella Concepta Goveas, Raja Selvaraj, and Ramesh Vinayagam

Abstract Arsenic is a tremendously harmful metal that exists naturally in the environment and is also discharged from anthropogenic sources. It is released into the air by high-temperature procedures such as volcanic eruptions, burning plants, flora, and coal-fired power plants. Recent studies have demonstrated the worldwide occurrence of arsenic in various ecosystems at elevated levels, culminating in contamination in foods and water. Chronic long-term exposure to arsenic causes cardiovascular, pulmonary, and renal diseases, unfortunate pregnancy consequences, newborn death, and cancers of different organ systems. This has resulted in a plethora of effective and viable physico-chemical and biological strategies for the removal of arsenic. Among all the removal methods, bioremediation, involving the use of microorganisms is the most efficient, owing to several advantages such as cost-effectiveness and environmental safety, with limited resources. In this context, this article presents a brief overview of the occurrence and toxicity of arsenic on human health in order to substantiate the importance of its removal from the polluted environment. Additionally, it also focuses on bioremediation of arsenic-contaminated samples by microorganisms such as bacteria, fungi and microalgae. These microorganisms are known to interact with arsenic by numerous mechanisms such as biosorption, redox reactions, methylation, and bioaccumulation. A detailed analysis of these mechanisms is provided along with the latest literature reports.

Keywords Arsenic · Toxicity · Bioremediation · Molecular mechanisms · Bacteria · Fungi · Microalgae

L. C. Goveas (✉)
Nitte (Deemed to be University), NMAM Institute of Technology (NMAMIT), Department of Biotechnology Engineering, Nitte, India
e-mail: goveas.louella@nitte.edu.in

R. Selvaraj · R. Vinayagam
Department of Chemical Engineering, Manipal Institute of Technology, Manipal Academy of Higher Education, Manipal, India

A. K. Yadav et al. (eds.), *Arsenic Removal Technologies*,
https://doi.org/10.1007/978-3-031-91171-2_1

1.1 Introduction

Amid the many nuisances faced by our environment, the most dominant is pollution, which is enhancing with time, culminating in a scenario of true crisis. Soil, water, and air are all exposed to hazardous pollutants by modernization, industrialization, and technical development. Among countless pollutants, such as petroleum hydrocarbons, plastics and microplastics, heavy metals, textile dyes, pesticides, and forever chemicals, arsenic, an innately existing metalloid, has instigated universal alarm, due to its worldwide occurrence and its role in producing substantial health hazards. Recent studies show that most of the environmental samples are contaminated with high levels of arsenic, and in about 70 countries, more than 140 million people are exposed to arsenic levels greater than 10 μg/L (limiting amount according to the World Health Organization) through drinking water (Frisbie and Mitchell 2022). According to the list of substances prioritized by the Toxic Substances and Disease Registry (ATSDR), which enlists harmful substances that pose a compelling threat to human health, arsenic is the first. It has also been classified as a Group 1 carcinogen by the International Agency for Research on Cancer (IARC) (Kumar et al. 2023). In addition to causing cancer of various organ systems, arsenic is linked with a multitude of well-documented complications and diseases in all organ systems of human body including effects on the fetus, newborns, and children.

To solve the concerns and difficulties associated with arsenic pollution, multiple treatment methodologies such as membrane filtration, advanced oxidation processes, ion exchange, adsorption coagulation/flocculation, and bioremediation have been utilized. Most of the methods listed are costly, generate secondary pollutants with limited efficiency, and are inconvenient for a large-scale application (Mahamallik and Swain 2023). The bioremediation of arsenic by microorganisms seems to be a convenient approach because of its low cost, sustainability, and environmental safety.

In the present era, the variety of microorganisms such as bacteria, fungi, microalgae, and their consortia have materialized as strategies inspired often by researchers to construct agents for arsenic bioremediation. These microorganisms are vital as they perform mechanisms such as biosorption, redox reactions, methylation, and bioaccumulation that aid in conversion of arsenic to less harmful forms (Ganie et al. 2024). The latest reports have accentuated the predominant role of microorganisms and their consortiums in reducing arsenic toxicity thereby eradicating the associated health risks, along with detailing their molecular mechanisms (Hashim 2023). In this regard, the present review provides a brief overview of the occurrence and toxic effects of arsenic on human health to accentuate the pressing need for its removal. It also elaborates the molecular mechanisms and microbial strategies for the bioremediation of arsenic from contaminated water and soil samples. A concise description of different types of microbes for the removal of arsenic is also highlighted, along with recent literature reports is also highlighted.

1.2 Arsenic: Occurrence and Toxic Effects on Human Health

Arsenic is found to be present in sulfidic minerals such as arsenopyrite, rock-forming minerals such as oxides, sulphides, phosphates, and silicates. Several other notable minerals containing arsenic include pyrites, orpiment, realgar, olivenite, arsenolite, and cobaltite, although sediments with high amounts of reducing agents also contain arsenic. Arsenic is discharged mostly into soil, water, and air by weathering, forest fires, volcanoes, and erosion of rocks. Solubilization of minerals with arsenic results in its liberation into the groundwater. Various human-caused activities, such as utilization of pesticides, electronics, glass and alloys, fertilizers, industrial wastes, wood, and fossil fuel burning, also release arsenic into the environment. It then eventually finds its way into plants, animals, and other living systems through consumption of nutrients from soil and drinking water (Nivetha et al. 2023; Shirin et al. 2021).

The mobility and toxicity of arsenic are predominantly dependent on its ionic forms. Arsenic can exist in pentavalent form (Arsenate, As^{5+}, As(V)) as mono or divalent ions under aerobic conditions, while in anaerobic, it exists in the trivalent (Arsenite, As^{3+}, As(III)), as anionic or uncharged species (Ganie et al. 2024). Additionally, organic forms of arsenic can occur under various methylated and oxidated forms such as arsenocholine, arsenobetaine, mono- and dimethylarsinic acid, mostly in mammalian systems. It has been confirmed that As (III) is potentially toxic than As(V), whereas organic arsenic is at least 100 times less toxic than its inorganic forms. As(V) is structurally similar to phosphate, it forms glucose-6-arsenate that resembles glucose-6-phosphate. This inhibits hexokinase and depletes ATP, instead of its generation. On the other hand, As (III) reacts with sulfhydryl and thiol groups, affecting enzymes and proteins throughout the human body (Genchi et al. 2022). Additionally, binding of As (III) with sulfhydryl groups leads to enhanced capillary permeability and dilation. Therefore, in most cases, mammalian systems try to reduce the tissue binding affinity of inorganic arsenic by methylating them into organic forms (Genchi et al. 2022).

Arsenic leads to three types of toxicity in humans, mainly based on the amount and duration of consumption, namely, acute, subacute, and chronic toxicity. Consumption of a significant amount of arsenic through contaminated food or water most often triggers acute and subacute arsenic poisoning, whose effects can be reversed with immediate medical attention. Within minutes to hours after consumption, nausea, projectile vomiting, profuse diarrhea, and severe abdominal pain are manifested, culminating in dehydration and hypotension. Few cases report neurological symptoms, hours after exposure such as headaches, seizures, and delirium (Hashim 2023). If arsenic exposure occurs by inhalation, respiratory symptoms such as cough, rapid breathing, and pulmonary oedema are noticed. Increased liver enzymes with liver necrosis may also be seen with acute arsenic toxicity. Subacute arsenic toxicity usually involves the manifestation of symptoms up to 1–3 weeks of exposure. Mostly, the nervous and hemotopoietic systems are affected with symptoms such as thrombocytopenia, hemolytic anemia, leukopenia, weakness, memory loss, seizures, hand

and foot tingling, and jaundice. Symptoms such as hemorrhage, proteinuria, and kidney failure may be seen days after ingestion (Ganie et al. 2024).

When small amounts of arsenic are consumed through contaminated water for a longer period of time, chronic toxicity is encountered. Chronic toxicity is known to affect all major organ systems such as reproductive, renal, respiratory, cardiovascular, nervous, gastrointestinal, hematopoietic, and skin manifestations (Hashim 2023) (Fig. 1.1). The high incidence of diabetes mellitus is observed in people with chronic arsenic exposure, while infant mortality, stunted growth, and abortions are also observed. Gastrointestinal and respiratory ingestion of arsenic causes cancers of the bladder, lung, skin, lung, liver, and blood.

Oral ingestion of arsenic in small amounts results in characteristic keratosis and melanosis on the skin. Various manifestations of melanosis include depigmentation, spotted pigmentation, hyper pigmentation, and leuco melanosis. After long-term exposure, spotted or nodular keratosis appears on the soles and palms of the individual, while arsenic starts to accumulate in the keratin-rich parts of the body, appearing as white lines in nails (Tang et al. 2023). Several studies depicted women exposed to arsenic experienced adverse pregnancy outcomes such as stillbirth, abortion, and pre-term birth at elevated rates compared to unexposed women. Furthermore, arsenic-exposed adolescents experienced delay onset of menstruation, hinting at endocrine dysfunction and reduced oestrogen secretion. Sperm production and

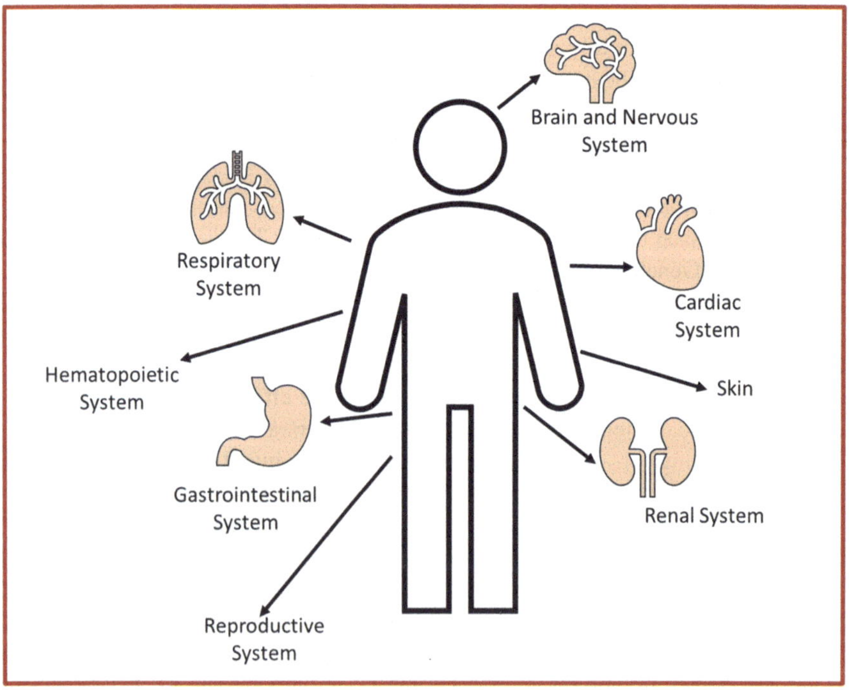

Fig. 1.1 Various systems of human organs affected by chronic exposure to arsenic

testosterone have also been observed in arsenic exposure in males (Garca-Vargas and Cebrián 2023). Severe impairment of the respiratory and cardiovascular systems has been observed due to chronic arsenic exposure, leading to a variety of disorders such as respiratory symptoms, chronic obstructive pulmonary disease, hypertension, ischemic heart problems, atherosclerosis, and mortality due to pulmonary and cardiac obstructions. Mood disorders, learning deficits, behavioral defects such as memory and movement, and impaired intellectual performance were noticed in children who encountered low levels of arsenic through diet or in utero. Most studies show limited effects of chronic exposure to arsenic on the renal system although in same cases tubular necrosis is induced in the same cases. The hematopoietic and immune system is intensely affected by chronic exposure to arsenic, causing anemia, thrombocytopenia, eosinophilia and immune suppression (Tang et al. 2023).

Arsenic consumption has been linked to cancers of the lung, prostate, liver, kidney, bladder, and skin, among which the most common is skin cancer, while lung cancer has the highest mortality (Tang et al. 2023). There are various mechanisms of cancer action of induction by arsenic exposure viz. gene amplification, oxidative stress, improper DNA repair and methylation, varied growth factors, suppression of p53, increased cell division and progression. Among skin cancers, the most common types are squamous cells, basal cells, and intraepithelial carcinomas. Prior to cancer manifestation, pre-cancerous lesions appear on the skin, such as keratosis and pigmentation, in most of the fatal cases (Kithmini and Perera 2024).

1.3 Microbial Molecular Mechanisms: Arsenic Remediation

Most of the molecular mechanisms aim to convert inorganic arsenic into organic forms, thereby reducing their toxicity. Heavy metals are detoxified and stabilized by microbial enzymes such as oxidoreductases through redox reactions. Several other microbial mechanisms of arsenic detoxification include biosorption, accumulation, and methylation (Wang et al. 2024) (Fig. 1.2).

Microorganisms oxidize potent As (III) into less toxic As(V) in the environment, catalysed by arsenite oxidase. The synthesized As(V) is further removed by precipitation with aluminium/iron oxides. As (V) can also be reduced to As (III) by microbial reduction during anaerobic respiration. This occurs through two distinct mechanisms: activation of arsenic reductases in the cytoplasm (Ars) and dissimilation by respiratory arsenic reductase (Arr). Biovolatilization, or biomethylation, refers to the microbial conversion of metals such as arsenic into a mixture of volatile and non-volatile methyl forms of metabolites. Aerobic bacteria frequently convert more toxic methylated arsenite into less toxic end products such as trimethylarsine oxide, along with release of trimethylarsine, a volatile gas. As (III) S-adenosylmethionine methyltransferase is the enzyme that catalyses this reaction via a seven-step process, where it transfers methyl groups from S-adenosylmethionine to As (III). For this

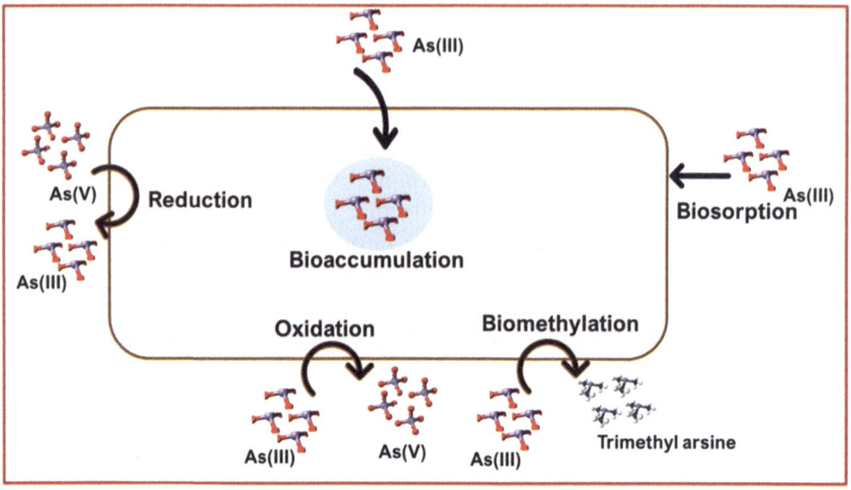

Fig. 1.2 Molecular mechanisms of arsenic removal by micro-organisms

reaction to occur, initially As (V) must be reduced to As (III) by two reduction reactions.

Biosorption and bioaccumulation are another two major processes in arsenic bioremediation. In biosorption, arsenic is removed via its attachment to the microbial cells, an energy-independent process, while in bioaccumulation, arsenic is up taken inside the microbial cell. The biosorption process involves physical and chemical adsorption, where there is interaction between the arsenic ions and the surface of the functional groups on the microbial cell. The functional groups may be sulfhydryl, phosphate, hydroxyl, amine, carbonyl, and carboxyl groups, while the interacting forces may be ionic, electrostatic, hydrophobic, Vander waal and π-π interactions. Additionally, the biosorption process is influenced by environmental parameters, microbial biomass, and arsenic properties. As biosorption is a metabolism-independent process, it works even with non-living microbial biomass. While living cells are used, they can accumulate arsenic intracellularly with subsequent transformation, after attachment to the cell wall (Wang et al. 2024). The transport of arsenic across the cell membrane depends on the difference in proton with chemical/electrical potential. This process depends on the metabolic strength of the cell, which eliminates arsenic through a variety of detoxification strategies. Unlike biosorption, bioaccumulation process depends on age and number of cells, in addition to culture and nutrient medium conditions (Kushwaha et al. 2023).

1.4 Micro-Organisms Involved in Arsenic Bioremediation

Micro-organisms such as bacteria, fungi, and microalgae are capable of removal and detoxification of arsenic by varied mechanisms (Sect. 3). They work by changing the valency of arsenic and enable its transformation or movement in the environment. Recent literature reports show that different species of bacteria, fungi, and microalgae are capable of remediating arsenic in contaminated samples (Table 1.1).

Bacteria interact with arsenic by oxidation, reduction, and methylation, while biosorption and accumulation are also frequently reported. As (III) oxidation serves as both catabolism and detoxification mechanisms for chemolithoautotrophic and heterotrophic bacteria. A chemolithotrophic *Delftia strain* tolerated and oxidised upto 70 mM As (III) with a removal % of 98 while different bacteria of *Bacillus, Shinella, Micrococcus,* and *Microbacterium*, although tolerated up to 500 mM As (V),

Table 1.1 Microorganisms involved in arsenic remediation

Microorganism	Source	Arsenic form	Removal mechanism	Efficiency (%)	References
Bacillus nealsonii ARP2	Paddy field	As(V)	Reduction	93.0 ± 0.2%	(Pandey et al. 2020)
Bacillus tequilensis ART2	Paddy field	As(III)	Oxidation	77.0 ± 0.14%	(Pandey et al. 2020)
Bacillus subtilis	Ground water	As(III)	Oxidation	88.74%	(Biswas and Sarkar 2019)
Bacillus flexus	–	As(III), As(V)	Biosorption	–	(Marwa et al. 2019)
Acinetobacter junii	–	As(III), As(V)	Biosorption	–	(Marwa et al. 2019)
Proteus alimentorum TY6	Soil	As(III), As(V)	Bioaccumulation	66.0 ± 0.471%, 96.0 ± 0.816%	(Kushwaha et al. 2023)
Pseudomonas aeruginosa K7Pb	Soil	As(III), As(V)	Bioaccumulation	78.0 ± 1.247%, 98.0 ± 0.942%	(Kushwaha et al. 2023)
Pseudomonas putida ARS1	Arsenic contaminated soil	As(III), As(V)	Reduction	91.0%	(Wang et al. 2024)
Chlorella vulgaris	–	As(V)	Biosorption	99.0%	(Islam et al. 2023)
Aspergillus niger	Soil	As(III)	Bioaccumulation	98.20%	(Duborská et al. 2020)
Pleurotus pulmonarius	Soil	As(III)	Bioaccumulation	–	(Zhang et al. 2023)
Phanerochaete chrysosporium	Acidic mine soil	–	Biosorption, Transformation	35.65%,	(He et al. 2022)
Miscanthus sinensis	Plant	–	Bioaccumulation	49.0%	(Yao and Zhou 2024)

were able to oxidize 8 mM As (III). Two strains of *Acinetobacter* and *Delftia* tolerated 6.8 g/L As (III) and removed it by a combined oxidation-adsorption approach with removal rates of 85–92% (Zhao et al. 2023). *Microvirga indica* S-MI1b, isolated from soil of the metal industry, completely oxidizes As (III) between pH 4.0 to 9.0 with an optimal 7.0. *Bosea* sp. AS-1, from the mine, tolerated 120 mM As(III) and oxidized 2 mM under heterotrophic conditions with yeast extract as a nitrogen source (Lu et al. 2018). An immobilized strain of *Pseudomonas migulae VC-19 represented 100%* oxidation of As (III) in 36 h. Some bacteria persist by using arsenic as a terminal electron acceptor. In addition to this, few others perform arsenic reduction, as they possess an arsenic reductase enzyme, which is known to offer resistance against arsenic. Strains of *Bacillus, Exiguobacterium,* and *Desulfitobacterium* resisted and reduced high amounts of As (V) and could also release arsenic from goethite. *The Bacillus* and *Pseudomonas* strains, from aquifer sediments, resisted As (III) and As (V), and reduced As (III) from various sources. Arsenic methylation and volatilization usually occur in soils contaminated by bacteria. A study carried out showed that *Pseudomonas alcaligenes rapidly* methylated As (III) into trimethyl arsenate (Zhang et al. 2015). Living bacteria accumulate arsenic intracellularly, while dead bacteria result in surface adsorption of arsenic. A strain of *Pseudomonas aeruginosa,* isolated from ground water removed arsenic by biosorption with 98% efficiency, while two bacterial strains adsorbed and accumulated As (III) along with the potential for its oxidation (Roy et al. 2024). Living cells of *Bacillus cereus* possessed exceptional ability for biosorption of As (III) from solutions (34.4 mg/g), whereas in another study, non-living cells of *Bacillus cereus,* isolated from a gold mine, also removed As (III) by biosorption with capacity (153.4 mg/g). *Acidithiobacillus ferrooxidans* BY-3, a chemolithotrophic bacteria capable of oxidising iron, also adsorbed As (III) and monomethyl arsonate in a spontaneous and feasible reaction (Zhang et al. 2015).

The use of fungi for the remediation of arsenic is gaining significance due to their dense mycelium, enzyme diversity, and rapid growth. Fungi can remediate arsenic by biosorption and detoxification through mechanisms including precipitation, sequestration, methylation, valency transformation, and active attachment or uptake (Duborská et al. 2020). The cell wall of fungi is rich in particular metal-attaching polysaccharides, peptides, and proteins with various functional groups such as amino, sulfate, phosphate, carboxyl, and hydroxyl groups, which are involved in mycoremediation of arsenic. Fungal strains belonging to *Aspergillus, Rhizopus, Trichoderma, Penicillium,* etc., have been often identified for removal of arsenic by adsorption and accumulation (Yao and Zhou 2024). A recent study conducted revealed that *Serendipita indica*, an endophytic fungus, upon its manifestation in the root of tomato plants, resulted in the accumulation of arsenic from the soil in the roots and prevented its transfer to other parts of the plant. The fungus immobilized arsenic in its cell wall and subsequently stored it within the vacuoles (Shukla et al. 2022). A strain of *Mucor circinelloides* also biosorbed arsenic on the cell wall and internalized it along with lead and cadmium from contaminated soil (Li et al. 2021). *Humicola* sp. 2WS1 isolated from arsenic-contaminated soil remediated arsenic by biomethylation and biovolatilization, while *Aspergillus* spp APR-1 and APR-2, isolated from industrial soil samples, immobilized in *Luffa aegyptiaca* biosorbed around 50% of arsenic from contaminated samples.

The detoxification of arsenic by algae involves a variety of metabolic pathways such as reduction of As(V) into As(III), As(III) oxidation, methylation, surface biosorption, and intracellular storage, while recent studies indicate arsenic transport by phosphate channels (Mao et al. 2022). *Chlorella vulgaris* adsorbed 91% of 50 mg/L As(III) from an aqueous solution in 76 min. As(III) attached to the cell wall of algae by interaction with functional groups such as hydroxyl, amino, amide, and carboxylate groups, as confirmed by FT-IR studies (Alharbi et al. 2023). A strain of *Chlorella pyrenoidosa,* on exposure to As(III) and As(V), accumulated both intracellularly. Additionally, HPLC studies confirmed the presence of monomethyl arsenic, which is responsible for the existence of additional redox and methylation reactions. This indicated the potential of *Chlorella pyrenoidosa* in subsequent arsenic detoxification, after accumulation (Zhang et al. 2020). Similarly, strains of *Leptolyngbya* sp. XZMQ and *Chlorella thermophila* SM01, identified from a mine, also adsorbed As(III) via interaction with hydroxyl, amine, and carboxyl groups on the cell surface. Then approximately 94% of As(III) was subsequently oxidised to As(V), confirming its detoxification (Mao et al. 2022).

1.5 Conclusions

One of the most promising and sustainable solutions for arsenic removal is bioremediation by using microorganisms. Bioremediation utilizes the inherent capabilities of bacteria, fungi, and algae to transform, absorb, accumulate, and detoxify arsenic, leading to its removal from polluted environment. Future prospects in this area include utilization of genetic engineering to influence the speed of remediation and facilitation of microbes to survive in unusual conditions. In addition, a microbial consortium could be applied to improve biosorption efficiencies along with the integration of these systems into suitable technologies.

Acknowledgements The authors express their heartfelt gratitude to our esteemed institute and all individuals who have played a role, directly or indirectly, in contributing to this study.

References

Alharbi, R. M., E. N. Sholkamy, K. I. Alsamhary, N. Abdel-Raouf, and I. B. M. Ibraheem, 2023. Optimization study of the capacity of *Chlorella vulgaris* as a potential bio-remediator for the bio-adsorption of Arsenic (Iii) from aquatic environments. *Toxics* 11(5):439–445. https://doi.org/10.3390/TOXICS11050439

Biswas, R., and A. Sarkar, 2019. Characterization of arsenite-oxidizing bacteria to decipher their role in arsenic bioremediation. *Preparative Biochemistry and Biotechnology* 49(1):30–37. https://doi.org/10.1080/10826068.2018.1476883

Duborská, E., K. Szabó, M. Bujdoš, H. Vojtková, P. Littera, E. Dobročka, H. Kim, and M. Urík, 2020. Assessment of *Aspergillus niger* strain's suitability for arsenate-contaminated water treatment and adsorbent recycling via bioextraction in a laboratory-scale experiment. *Microorganisms* 8(11):1668–1679. https://doi.org/10.3390/MICROORGANISMS8111668

S. J. S. Flora, 2015. *Handbook of Arsenic Toxicology*, Ist ed. Academic Press https://doi.org/10.1016/B978-0-323-89847-8.00024-9

Frisbie, S. H., and E.J. Mitchell, 2022. Arsenic in drinking water: An analysis of global drinking water regulations and recommendations for updates to protect public health. *Plos One*, 17(4):125–130. https://doi.org/10.1371/JOURNAL.PONE.0263505

Ganie, S. Y., D. Javaid, Y.A. Hajam, and M. S. Reshi, 2024. Arsenic toxicity: Sources, pathophysiology and mechanism. *Toxicology Research* 13(1):1–10 https://doi.org/10.1093/TOXRES/TFAD111

García-Vargas, G. G., and M. E. Cebrián, 2023. *Health Effects of Arsenic*. Edited by L. W. Chang, *Toxicology Metals Volume I,* CRC Press, https://doi.org/10.1201/9781003418917

Genchi, G., G. Lauria, A. Catalano, A. Carocci, and M. S. Sinicropi, 2022. Arsenic: a review on a great health issue worldwide. *Applied Sciences* (12):6184–6195. https://doi.org/10.3390/APP12126184

Hashim, A. 2023. Human Health Effects of Chronic Arsenic Exposure. *Environmental Science and Engineering, Part F* 1324: 45–60. https://doi.org/10.1007/978-3-031-37561-3_3

He, N., L. Hu, C. Jiang, and M. Li, 2022. Remediation of chromium, zinc, arsenic, lead and antimony contaminated acidic mine soil based on *Phanerochaete chrysosporium* induced phosphate precipitation. *Science of the Total Environment* 850:157995. https://doi.org/10.1016/J.SCITOTENV.2022.157995

Islam, M. S., I. Maamoun, O. Falyouna, O. Eljamal, and B. B. Saha, 2023. Arsenic removal from contaminated water utilizing novel green composite *Chlorella vulgaris* and nano zero-valent iron. *Journal of Molecular Liquids* 370:121005. https://doi.org/10.1016/J.MOLLIQ.2022.121005

Kithmini, N. K. S., and O. D. A. N. Perera, 2024. Arsenic Exposure Through Dietary Intake and Associated Health Hazards In *Arsenic Toxicity Remediation* edited by N. Kumar, M. Z. Hashmi and S. Wang, Cham: Springer Nature Switzerland https://doi.org/10.1007/978-3-031-52614-5_5

Kumar, A., M. Ali, V. Raj, A. Kumari, M. Rachamalla, S. Niyogi, D. Kumar, A. Sharma, A. Saxena, G. Panjawani, P. Jain, A. Vidyarthi, N. Kumar, M. Kumar, P. K. Niraj, M. Rahman, A. Bishwapriya, R. Kumar, M. Sakamoto and A. K. Ghosh, 2023. Arsenic causing gallbladder cancer disease in Bihar. *Scientific Reports* 13(1):1–16. https://doi.org/10.1038/s41598-023-30898-0

Kushwaha, R., R. S. Singh, and D. Mohan, 2023. Arsenic resistance and accumulation by two bacteria isolated from Ratanpur village of Varanasi district, Uttar Pradesh. *Bioresource Technology Reports,* 24:101639. https://doi.org/10.1016/J.BITEB.2023.101639

Li, X., X. Lan, X. Feng, X. Luan, X. Cao, and Z. Cui, 2021. Biosorption capacity of *Mucor circinelloides* bioaugmented with *Solanum nigrum* L. for the cleanup of lead, cadmium and arsenic. *Ecotoxicology and Environmental Safety* 212:112014. https://doi.org/10.1016/J.ECOENV.2021.112014

Lu, X., Y. Zhang, C. Liu, M. Wu, and H. Wang, 2018. Characterization of the antimonite- and arsenite-oxidizing bacterium *Bosea* sp. AS-1 and its potential application in arsenic removal. *Journal of Hazardous Materials* 359: 527–534. https://doi.org/10.1016/J.JHAZMAT.2018.07.112

Mahamallik, P., and R. Swain, 2023. A mini-review on arsenic remediation techniques from water and future trends. *Water Science and Technology* 87(12): 3108–3123. https://doi.org/10.2166/WST.2023.190/1245840/WST087123108.PDF

Mao, Q., Z. Xie, S. Irshad, Z. Zhong, T. Liu, F. Pei, B. Gao, and L. Li, 2022. Effect of arsenic accumulation on growth and antioxidant defense system of *Chlorella thermophila* SM01 and

Leptolyngbya sp. XZMQ. *Algal Research* 66:102762. https://doi.org/10.1016/J.ALGAL.2022.102762

Marwa, N., N. Singh, S. Srivastava, G. Saxena, V. Pandey, and N. Singh, 2019. Characterizing the hypertolerance potential of two indigenous bacterial strains (*Bacillus flexus* and *Acinetobacter junii*) and their efficacy in arsenic bioremediation. *Journal of Applied Microbiology* 126(4):1117–1127. https://doi.org/10.1111/JAM.14179

Nivetha, N., B. Srivarshine, B. Sowmya, M. Rajendiran, P. Saravanan, R. Rajeshkannan, M. Rajasimman, T. H. T. Pham, V. K. Shanmugam, and E. N. Dragoi, 2023. A comprehensive review on bio-stimulation and bio-enhancement towards remediation of heavy metals degeneration. *Chemosphere* 312: 137099. https://doi.org/10.1016/J.CHEMOSPHERE.2022.137099

Pandey, N., K. Manjunath, and K. Sahu, 2020. Screening of plant growth promoting attributes and arsenic remediation efficacy of bacteria isolated from agricultural soils of Chhattisgarh. *Archives of Microbiology* 202(3):567–578. https://doi.org/10.1007/S00203-019-01773-2/METRICS

Roy, V., B. K. Saha, S. Adhikary, M. G. Chaki, M. Sarkar, and A. Pal, 2024. Isolation, characterization, identification, genomics and analyses of bioaccumulation and biosorption potential of two arsenic-resistant bacteria obtained from natural environments. *Scientific Reports* 14(1):1–16. https://doi.org/10.1038/s41598-024-56082-6

Shirin, S., & Yadav, A. K. (2021). Assessment and suitability for the dumping of flyash in an abandoned mine. *Environmental Challenges*, 4, 100111. https://doi.org/10.1016/j.envc.2021.100111

Shukla, J., S. Mohd, A. S. Kushwaha, S. Narayan, P. N. Saxena, L. Bahadur, A. Mishra, P.A. Shirke, and M. Kumar, 2022. Endophytic fungus *Serendipita indica reduces arsenic mobilization from root to fruit in colonized tomato plant. Environmental Pollution* 298: 118830. https://doi.org/10.1016/J.ENVPOL.2022.118830

Tang, G. T., J. Elakis, and L. Scardamaglia, 2023. Cutaneous manifestations and treatment of arsenic toxicity: A systematic review. *Skin Health and Disease* 3(4):e231. https://doi.org/10.1002/SKI2.231

Wang, Z. W., G. Yang, J. Chen, Y. Zhou, A. Núñez Delgado, H. L. Cui, G. L. Duan, B. P. Rosen, and Y. G. Zhu, 2024. Fundamentals and application in phytoremediation of an efficient arsenate reducing bacterium *Pseudomonas putida* ARS1. *Journal of Environmental Sciences* 137: 237–244. https://doi.org/10.1016/J.JES.2023.02.020

Yao, S., and B. Zhou, 2024. Enhancing phytoremediation of cadmium and arsenic in alkaline soil by *Miscanthus sinensis*: A study on the synergistic effect of endophytic fungi and biochar. *Science of The Total Environment* 923:171458. https://doi.org/10.1016/J.SCITOTENV.2024.171458

Zhang, J., T. Cao, Z. Tang, Q. Shen, B. P. Rosen, and F. J. Zhao, 2015. Arsenic methylation and volatilization by arsenite s-adenosylmethionine methyltransferase in *Pseudomonas alcaligenes* NBRC14159. *Applied and Environmental Microbiology*, 81(8): 2852–2860. https://doi.org/10.1128/AEM.03804-14/SUPPL_FILE/ZAM999116186SO1.PDF

Zhang, J, F. Zhou, Y. Liu, F. Huang, and Zhang, C. 2020. Effect of extracellular polymeric substances on arsenic accumulation in *Chlorella pyrenoidosa. Science of The Total Environment* 704: 135368. https://doi.org/10.1016/J.SCITOTENV.2019.135368

Zhang, Y., X. Chen, and L. Xie, 2023. Pleurotus pulmonarius Strain: Arsenic(III)/Cadmium(II) accumulation, tolerance, and simulation application in environmental remediation. *International Journal of Environmental Research and Public Health* 20(6):5056. https://doi.org/10.3390/IJERPH20065056

Zhao, M., G. Zheng, X. Kang, X. Zhang, J. Guo, M. Zhang, J. Zhang, Y. Chen, and L. Xue, 2023. Arsenic pollution remediation mechanism and preliminary application of arsenic-oxidizing bacteria isolated from industrial wastewater. *Environmental Pollution* 324:121384. https://doi.org/10.1016/J.ENVPOL.2023.121384

Chapter 2
Health Impacts of Arsenic Exposure and Regulations for Safe Drinking Water

Praveen Kumar, Muthukumar Raghunathan, and Ashish Kapoor

Abstract Arsenic (As) occurrence in aqueous systems is a critical global health concern, affecting a vast population, particularly in areas dependent on groundwater sources. Persistent exposure to As for an extended period, even at trace level concentrations, is associated with numerous health issues affecting various organs and systems, including the cardiovascular, endocrine, immune, nervous, and respiratory systems. The cumulative nature of arsenic exposure intensifies the health risks, taking years or even decades for the manifestation of diseases. This chapter aims to elucidate the health-related concerns of arsenic toxicity, focusing on the factors that influence individual susceptibility. It also reviews global regulatory guidelines regarding arsenic concentration in drinking water. The need to enforce these regulations is thereby emphasized and the importance of continuous monitoring and preventive and control measures to prevent arsenic-related health issues is highlighted. Overall, the chapter discusses the urgent necessity for stringent regulations and effective public health measures in the context of arsenic exposure to ensure the availability of safe potable water and public well-being.

Keywords Arsenic · Health effects · Contamination · Drinking water · Guidelines

2.1 Introduction

Arsenic (As) is an element that occurs naturally in the environment and is ubiquitously present in various ecological matrices. Its mobilization pathways include natural phenomena such as weathering reactions, volcanic emissions in conjunction with human-induced factors, such as fossil-based fuel combustion, mining, and use of

P. Kumar · M. Raghunathan
Department of Biotechnology, School of Bioengineering, SRM Institute of Science and Technology, Kattankulathur, Chennai, Tamil Nadu, India

A. Kapoor (✉)
Department of Chemical Engineering, Harcourt Butler Technical University, Kanpur, Uttar Pradesh, India
e-mail: ashishk@hbtu.ac.in

A. K. Yadav et al. (eds.), *Arsenic Removal Technologies*,
https://doi.org/10.1007/978-3-031-91171-2_2

herbicides and desiccants. The contamination of As in aqueous systems eventually leads to its entry into the food chain and poses grave risks to the health of living beings. Various common sources of arsenic exposure are represented in Fig. 2.1.

As in potable water remains a grave concern due to its hazardous impacts. Chronic exposure to As-contaminated water may lead to several diseases. Water quality regulations around the globe have been tightened to control arsenic levels in potable water due to health risks linked to this element (Ali et al. 2019; Jha et al. 2017). To comply with these regulations, water systems must be tested and monitored often. In addition, sources of arsenic must be remedied and treated using advanced filtration equipment. Educating people about the risks of arsenic and how to minimize their exposure is another crucial aspect of maintaining clean drinking water. By strictly adhering to these guidelines and continually improving our water treatment processes, the risks of arsenic exposure can be reduced, and accessibility to safe drinking water can be enhanced (Mitchell et al. 2011; Shirin et al. 2014). In this perspective, the aim of this chapter is to proffer an overall view of key health aspects related to the hazardous nature of As and regulatory guidelines for drinking water.

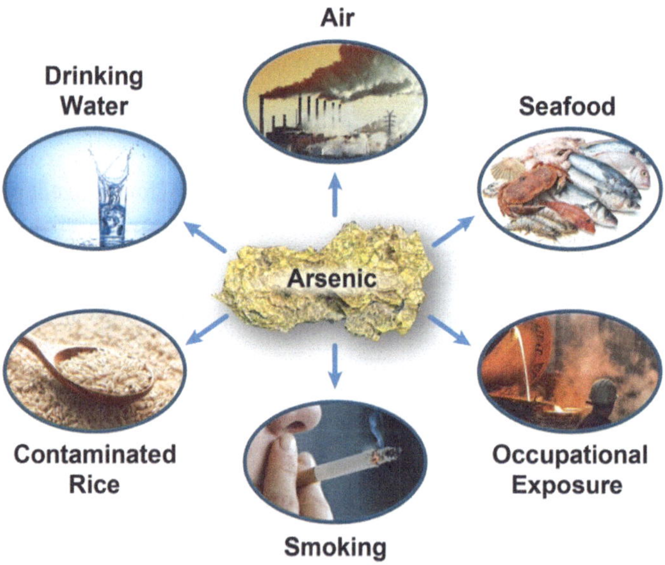

Fig. 2.1 Common sources of arsenic intake for the general population (reprinted from Liu et al. (2023) under CC-BY license)

2.2 Health-Related Implications of Exposure to As

The dosage and exposure time of As are pivotal in determining the extent to which it may harm health. Arsenic has the ability to cause numerous health complications in humans, including problems with reproduction, the nervous system, the heart, the liver, cancer, and diabetes. Figure 2.2 illustrates the adverse health outcomes associated with As toxicity. The key effects on human health are explored in the ensuing subsections.

2.2.1 Dermatological Effects

Skin cancer, Bowen's disease, keratosis, melanosis, and leucomelanosis are among the many skin diseases and disorder that arsenic is known to induce. Several studies have been conducted to examine these aspects. A notable research study was conducted to evaluate the correlation between levels of As in consumable water and the prevalence of arsenical dermatosis in specific Mongolian regions. About 96% of the water from the tubewells in Wuyuan and nearly 69% in Alashan had As concentration levels exceeding 50 μg/L. The prevalence of arsenical dermatosis was 44.8% among 1,176 patients in Wuyuan and 37.1% among 433 patients in Alashan. The study found that arsenical dermatosis was more common in adults over the age of 40 years (Guo et al. 2001). Chronic consumption of arsenic at a daily dose of 10 μg/kg has been associated with dermatosis and other forms of arsenic poisoning.

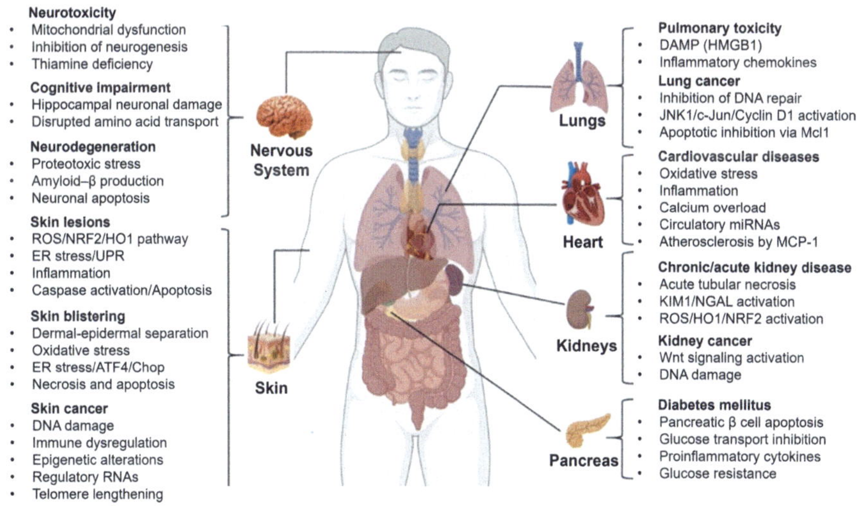

Fig. 2.2 Health impacts of arsenic exposure (adapted from Muzaffar et al. (2023) under CC-BY license)

The reported studies have observed a discernible connection of As with skin, kidney, bladder, and pulmonary issues. Water having As levels exceeding 50 μg/L augments the risks of lung and bladder malignancies, according to researchers. Furthermore, as can be shown in Table 2.1, several studies have connected arsenic exposure through water consumption with skin lesions in arsenic-affected regions in Bangladesh and India. Study participants had skin lesions in four regions of Bangladesh where arsenic was detected in the water supply of the 1,481 people surveyed, 430 had arsenical skin lesions. The concentration of As in water had values from undetectable to 2,040 μg/L. After adjusting for age, the proportion of skin problems increased from 18.6% in lowest exposure group with As concentrations of 150 μg/L or less to 37% in the highest exposure group with concentrations greater than 1,000 μg/L for men, and to 24.9% for women. Researchers found that in 14 villages in West Bengal, arsenic dermatosis was related to the presence of As in drinking water, which was around 200 μg/L. Skin problems were also observed to be caused by arsenic (As) in 7,683 people in West Bengal who drank water from areas polluted with metal. Age-adjusted prevalence figures show that hyperpigmentation occurred at a rate of 0.3% in the group with the lowest exposure (<50 μg/L). In the group exposed to the highest concentrations (≥800 μg/L), the rates increased to 8.3% and 11.5%, respectively. After accounting for age, studies conducted by (Guha Mazumder et al. 1998 and Rahman et al. 2001) revealed that in the studied group (<50 μg/L), the risk of keratosis was 0.2%, while in the other group (≥800 μg/L), it was 10.7% and 22.7% per 100 persons, respectively.

Table 2.1 Studies illustrating the relation between arsenic concentration levels and manifestation of health issues

Region/Country	Health effect	Concentration of As (μg/L)	References
West Bengal, India	Skin lesions	300	Chakraborti et al. (2004)
China	Skin lesions	50	Guo et al. (2001)
Inner Mongolia	Keratosis and pigment disorder	50	Guo et al. (2006)
Cordoba, Argentina	Recognizable signs of As toxicity	100	Astolfi et al. (1981)
West Bengal, India	Skin lesions (keratosis and hyperpigmentation)	50	Guha Mazumder et al. (1998)
Bangladesh	Skin lesions (keratosis and melanosis)	150	Tondel et al. (1999)
Bangladesh	Skin lesions (keratosis and melanosis)	50	Ahsan et al. (2000)

2.2.2 Diabetes Mellitus

Studies have shown that the resident population in the localities of Taiwan where blackfoot disease is a common occurrence have a higher frequency of diabetes mellitus. Moreover, it is also found that excessive levels of As in water are associated with this condition. People in these southern Taiwanese districts who were exposed to toxic As due to consuming water with As levels (range: 250 to 1140 μg/L, median: 780 μg/L) were likely to have a critical health risk from diabetes mellitus. Tseng et al. noted that people who consumed arsenic-contaminated well water at values between 700 and 930 μg/L were more likely to develop diabetes (Tseng et al. 2000). A community from an As-endemic area in the north Mexican state of Coahuila was studied to determine the prevalence of type 2 diabetes and the relation between elevated As concentration in potable water (in a range of 20 to 400 μg/L) (Coronado Gonzalez et al. 2007). As urine concentrations were shown to be dose-dependently related to type 2 diabetes mellitus, according to their research.

2.2.3 Hepatic Effects

Researchers in West Bengal areas where arsenic-contaminated water existed observed liver damage, with amounts ranging 200–2000 μg/L (Guha Mazumder et al. 1998). Hepatomegaly was found in 92.6% of those exposed to As exposure levels above 50 μg/L, compared to 6.2% in those with lower levels. In another study, liver function was investigated in three communities in the Lagunera area of Mexico. The drinking water in these communities had an average arsenic concentration ranging from 14 to 300 μg/L. Serum levels of alkaline phosphatase and bilirubin showed marked elevation where arsenic exposure levels were significantly high (Hernandez-Zavala et al. 1998). Among ninety-three people in West Bengal, who had severe hepatomegaly because of persistent arsenicosis, neither bilirubin nor alkaline phosphatase levels were significantly elevated. According to (Santra et al. 1999), out of 69 cases of chronic poisoning, 63 (91.3%) developed portal fibrosis, 2 (2.9%) developed cirrhosis and 4 (5.8%) showed normal histology.

2.2.4 Pregnancy Effects

Investigative studies have indicated that women subjected to a high arsenic content in potable water have a higher risk of spontaneous miscarriages. Portable water consumption with As concentrations greater than 100 μg/L has been associated with an increase in the appearance of unprompted abortions, stillbirths, and neonatal deaths. The same was explored in a study of Bangladeshi women when arsenic concentrations went over 240 μg/L (von Ehrenstein et al. 2006). Women who were exposed to arsenic throughout their pregnancies had elevated levels of the harmful metal in their placentas and umbilical cords. Arsenic was detected in the placentas and cord blood of mothers exposed to a concentration of 200 μg/L As concentration, in accordance with the research carried out in Argentina. Arsenic-contaminated drinking water with levels between 90 and 860 μg/L was associated with higher prenatal and neonatal mortality rates and lower birth weight, according to studies carried out in northern Chile. Premature births occurred at a slightly higher rate in a Taiwanese town where the arsenic concentrations in potable water varied from below the detectable value to 3590 μg/L (Yang et al.2003). Another study indicated that 533 pregnant women who consumed As-contaminated water were prone to spontaneous miscarriage or stillbirth (Milton et al. 2005). Researchers also found that pregnant women in the Bangladeshi village of Eruani had a variety of difficulties throughout their pregnancies, including premature deliveries, stillbirths, and spontaneous abortions. As range of 201 to 1,200 μg/L was detected in the exposed group. There appeared to be a dose and response relationship between risk of newborn mortality and the mean arsenic amount in potable water (Ahamed et al. 2001). There were major problems for pregnant women in Uttar Pradesh, Bihar, and West Bengal whose water came from areas where arsenic was present. Arsenic exposure was associated with higher rates of adverse pregnancy outcomes, including stillbirth, premature birth, low birth weight, and infant mortality, compared to population in noncontaminated locales (Rahman et al. 2007). In another study, an increased chance of stillbirth was associated with elevated levels of As exposure during pregnancy, but no link appeared to manifest with spontaneous abortion or neonatal death (von Ehrenstein et al. 2006).

2.2.5 Cardiovascular Effects

Consumption of arsenic increases the incidence of hypertension and death associated with cardiovascular disease, according to epidemiological research. According to the World Health Organization (WHO), peripheral vascular disease is more common in countries where arsenic is present in water for extended periods, including Mexico, Chile, the United States, and Taiwan. A possible link between low As levels and an increased likelihood of hypertension was identified in research carried out in Michigan. In Antofagasta, Chile, researchers compared cardiovascular symptoms between 144 participants who had arsenical skin lesions and 36 participants who did not have lesions. According to Wang et al. (2007), the possibility of a relation between As exposure and cardiovascular mortality was examined in thirty US counties with a mean arsenic amount in potable water greater than 5 μg/L. The study looked at 8,102 men and women from 3,901 households in Taiwan to see whether there was a connection between the composition of As in water and the occurrence of cerebrovascular disease. After taking into account demographic and health-related factors such as age, sex, blood pressure, diabetes, smoking, and alcohol use, a significant association was identified between arsenic levels in water and the incidence of cerebrovascular disease.

Chronic arsenic toxicity is also likely to contribute to the development of hypertension (high blood pressure) in humans. There were 1,481 cases of hypertension in regions of Bangladesh where arsenic was detected, compared to 114 cases in areas where the element was not present. In 582 individuals from southern Taiwanese regions known to be arsenic endemic and who had refrained from drinking polluted well water for more than two decades, researchers examined the relationship of arsenic exposure with peripheral vascular disease. Various factors, such as age, gender, body mass index, smoking habits, and long-term exposure to arsenic, were taken into account when investigating the dose and response relationship. From 1971 to 1994, Tsai et al. examined the mortality rates due to peripheral vascular disease to national and local rates in Taiwan's arsenic-endemic districts and counties (Tsai et al. 2003). Compared to non-endemic areas, the mortality rates for both men and females in Taiwan's endemic regions were much higher due to vascular disease (Tseng et al. 1996).

2.2.6 *Respiratory Effects*

Respiratory problems such as coughing, chest pain, bronchitis, and difficulty breathing may develop after ingesting inorganic compounds over a prolonged period. In the reported studies, the influence of As borne on potable water was investigated, as was frequency of respiratory symptoms such as coughing, chest noises, and difficulty breathing in the 24-South Parganas area of West Bengal, which is affected by pollution. As levels of 7,683 people, ranging in age from 9 to 60, varied from 3 to 3,400 μg/L. According to (Guha Mazumder et al. 2000), while controlling for age, the male prevalence odds ratio (POR) in case of coughing was 5 and the female POR was 7.8, while the male POR for chest sounds was 6.9 and the female POR was 9.6. Among males, POR for difficulty breathing was 3.7, whereas among women, it was 23.2. In order to determine if arsenic exposure causes chronic bronchitis, cross-sectional research in Bangladesh included 94 people (ranging in age from 5 to 51 years) who had arsenical skin lesions (Milton and Rahman 2002). Their potable water had As levels between 136 and 1,000 μg/L. Chronic bronchitis symptoms were seen in fifteen of the fifty-four women and fourteen of the forty males in the exposed group. According to (Milton et al. 2003), the unhealthy condition was found in 2 out of 74 women and 11 out of 50 males in the non-exposed group.

2.2.7 Carcinogenic Effects

Epidemiological studies provide substantial evidence that inorganic As has a carcinogenic nature. The link between As exposure and various types of cancer has been an area of intensive research. According to other reported studies, lifetime risk of cancer mortality could approach 13 per 1,000 people if they drank 1L of water with 50 μg/L of As every day (Smith et al. 2002). A cumulative cancer mortality risk of 1 in 100 might be experienced by those subjected to As exposure of 50 μg/L concentration. Furthermore, the use of water with As levels over 100 μg/L was associated with a substantial concern about skin cancer and arsenic poisoning, according to studies conducted in Córdoba, northern Argentina. In a cohort study that included 113 individuals, the researchers found a marked increase in the risks of cancer mortality in those subjected to arsenic concentrations greater than 1000 μg/L. The standardized mortality rate for lung cancer was 15.69. The chance of developing uterine and liver cancers was also shown to be higher. The odds ratios for men were 5.31 and for women 12.01 in southwest Taiwan, where As concentration values in potable water exceeded 600 μg/L, in comparison to those subjected to As levels below 290 μg/L. In addition, researchers discovered that among men and women in Taiwan, the types of liver, bladder, lung, and kidney cancer were the leading causes of death. The dose and response relationships were also discovered in a study that demonstrated that As toxicity in water was associated with an elevated incidence of renal and lung cancers. The standard mortality rates for kidney cancer were 0.87 to 1.57 for men and 1.00 to 1.81 for women. For lung cancer, the ratios were 0.92 to 1.77 for men and 1.24 to 2.16 for women. The impact of arsenic levels on the incidence of cancer was also investigated (Chen et al. 1995).

2.2.8 Neurological Effects

Peripheral neuropathy could be the result of prolonged As exposure at concentrations between 100 and 1,000 μg/L. A few patients from West Bengal, India, who suffered from arsenical dermatoses and had As exposure levels between 200 and 2,000 μg/L, were found to have sensory predominance distal polyneuropathy. Populations in some Indian states have been associated with neurological problems due to As poisoning originating from water consumption. The study's authors inferred that As in groundwater in West Bengal and other parts might cause nervous system-related problems. Most people who are poisoned with arsenic have peripheral neuropathy. Neuropathy is often chronic but may be subacute. In highly polluted regions of some countries, where As concentrations in water supplies ranged between 300 and 1,584 μg/L, neurological problems were observed in individuals with arsenicosis owing to long-term exposure to the metal. As concentration values ranged up to 3,590 μg/L in a research study conducted in Taiwan, suggesting that neurobehavioral consequences might occur during adolescence as a result of persistent exposure (Tsai et al. 2003).

The cognitive capacities of young people are likely to be negatively affected due to arsenic toxicity. Children whose exposure to arsenic levels beyond 50 μg/L had worse performance than those whose exposure was below 5.5 μg/L, suggesting that this association was dose-dependent (Wasserman et al. 2004). Neurotoxicity symptoms, such as impaired hearing, taste, vision, tingling, and numbness, have recently been shown to be much more common among arsenic-affected people in Mongolia (Guo et al. 2001).

Arsenic toxicity has several detrimental effects on cells, as depicted in Fig. 2.3. It induces apoptosis, leading to programmed cell death. Additionally, arsenic facilitates the production of reactive oxygen species, resulting in oxidative stress and cellular damage. It also inhibits DNA repair, resulting in genomic instability and potential mutations. Furthermore, arsenic can cause post-translational modifications that alter protein functions and signaling pathways. In addition, it leads to DNA hypomethylation and cell proliferation, which can contribute to carcinogenesis.

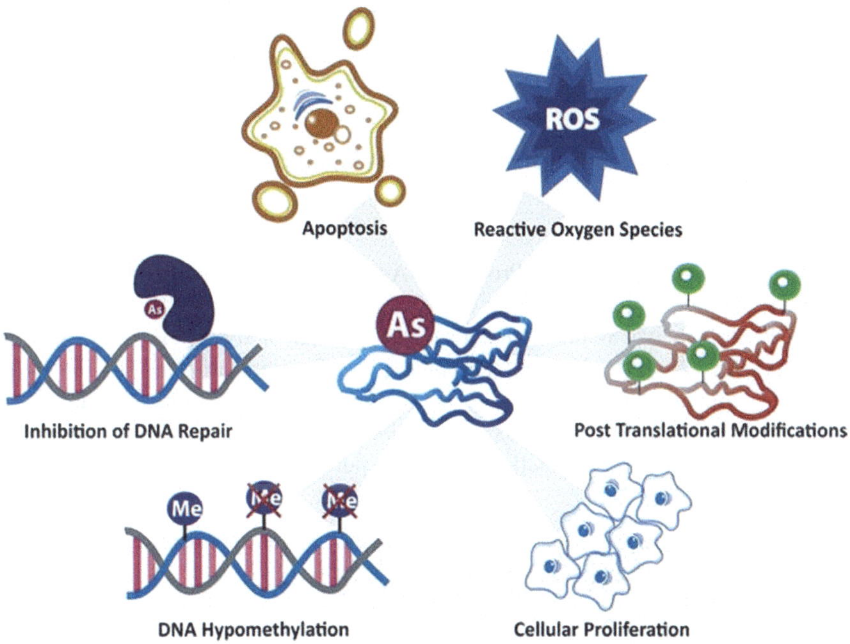

Fig. 2.3 Impacts of arsenic at cellular level (reprinted from Hussain et al. (2018) under CC-BY license)

2.3 Arsenic Exposure Regulations

The majority of nations have passed legislation establishing maximum levels of As that may be legally found in drinking water. However, the regulations in some low- and middle-income nations continue to be based on the 1963 standards set by WHO, which allow a concentration of 50 μg/L. In contrast, most high-income nations follow the updated 1993 standard of a threshold level As level of 10 μg/L. Standards must be reviewed periodically and kept up-to-date with the latest guidelines (WHO 2021). Figure 2.4 illustrates the global scenario of drinking water standards for arsenic for 180 countries.

The total allowed concentration is 10 μg/L in 134 countries, less than 10 μg/L in 2 nations, and more than 10 μg/L in 40 countries. As in drinking water, restrictions have been in existence since 1985, with the most current one taking effect in 2021. The most common year for regulatory changes was 2011, with 2017 somewhat in the centre. The guidelines are set by regulatory bodies, including 85 nations with laws or decrees, 58 ministries or organizations, and 33 national standards boards (Frisbie and Mitchell 2022).

2.3.1 *Guidelines for Drinking Water Based on Health Standards*

In 2000, the US Environmental Protection Agency (EPA) recommended a health-based goal, known as the "Maximum Contaminant Level Goal" (MCLG) for arsenic at 0 μg/L, though it was not mandatory. The MCLG of 0 μg/L is still valid. In 1999, the US EPA established the "Practical Quantification Limit" (PCL) for As at 3 μg/L. A national standard of 5 μg/L was projected after a cost–benefit analysis, instead of using PQL being used as the mandatory "maximum contaminants level" (MCL). Despite the early opposition of the US, public input led to an increase from

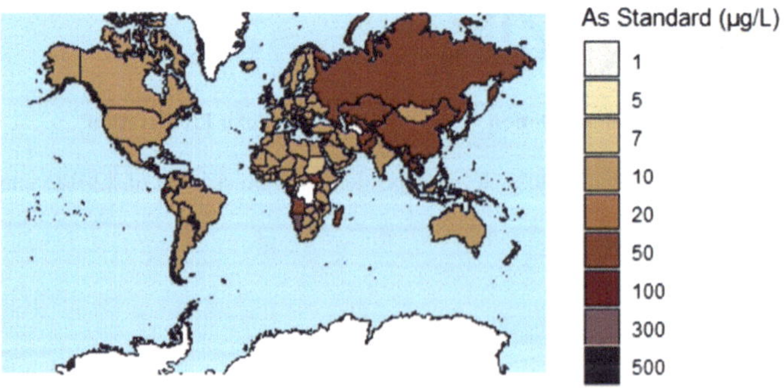

Fig. 2.4 World map indicating arsenic (As) standards for various nations (reprinted from Frisbie and Mitchell (2022) under CC-BY license)

5 μg/L to 10 μg/L, and this level was ultimately established as the national norm. The suggestion for a lower limit by the EPA Scientific Advisory Board is also being pursued. The California Environmental Protection Agency (CalEPA) has established a "Public Health Target" (PHG) of 0.004 μg/L for potable water. In this PHG, which is focused on the effects on human health, the detection limits and water treatment systems are not taken into account. A PHG must have a low potential for long-term negative health effects and be supported by current scientific evidence in order to be legal in California. The PHG level of 0.004 μg/L is 2500 times below the US EPA standard of 10 μg/L for potable water, as the well as provisional guideline recommended by WHO.

2.3.2 Regulatory Drinking Water (DW) Guidelines

In 1958, the WHO formulated guidelines for drinking water to be 200 μg/L to address health issues. The limit of occurrence of arsenic in potable water was reduced to 50 μg/L in 1963 after more research on the undesirable impacts of As on public health. While some epidemiological studies had indicated that arsenic may be carcinogenic, there was no conclusive evidence to confirm its carcinogenic effects on humans through portable water. After noting that, the WHO classified the 50 μg/L recommendation as a tentative limit in 1971. It is however advisable to keep the As concentration in potable water at the lowest possible value. Due to the absence of practical, cost-effective, sensitive, and dependable methods for regularly analyse As levels in drinking water, a threshold of 50 μg/L was considered as low as possible in 1971. Furthermore, interferences and inadequate reduction in analyte mean that the polarographic estimation approach may often be erroneous. Due to the partial formation of AsH_3 from the sample matrix, the accuracy of the silver diethyldithiocarbamate spectrophotometric approach for measuring As may be compromised. Atomic absorption spectroscopy (AAS) is a great alternative, but it is too expensive for many developing countries to use despite its sensitivity, accuracy, and precision.

The World Health Organization shifted from providing "standards" to issuing "guidelines" for drinking water in 1984. National authorities and stakeholders should use guidelines as a guide; however, they should not be seen as global standards. Officials at the federal level are required to create guidelines taking into account regional circumstances, resources, and risks. The prior WHO standard for As in drinking water, 50 μg/L, was maintained in the 1984 WHO guideline. A provisional guideline of 10 μg/L for arsenic in potable water was established in 1993 by the WHO, replacing the previous tentative limit of 50 μg/L. The danger of skin cancer had been well publicized at this stage, and suspicions of other malignancies had begun to circulate. As far as human carcinogens go, inorganic As is in Group 1 according to the International Agency for Research on Cancer. In cases where there is an adequate indication of carcinogenicity in humans, this category is used. There is evidence that populations that use arsenic-contaminated water have an elevated risk of skin cancer and maybe other malignancies. This risk grows with age and dose.

The WHO has put the value on the basis of health at 0.17 μg/L, which is lower than the practical quantitative limit of 10 μg/L. This recommendation was considered provisional due to the analytical feasibility. Instead of considering health data, the practical analytical actors that informed the 1993 WHO drinking water recommendation for As at 10 μg/L led to a lower limit. Skin, bladder, and lung cancers are among the many types of cancer that are often associated to higher concentration of As in potable water, according to epidemiological research. All successive versions and additions of the Guidelines for Drinking Water Quality since 1993 have upheld the provisional level of 10 μg/L for As in potable water, formulated by the WHO. In 2011, the World Health Organization expanded this list to include kidney carcinoma. An increased risk of various forms of cancers, together with other skin abnormalities including hyperkeratosis and pigmentation changes, is associated with prolonged exposure to arsenic in potable water, according to the International Program on Chemical Safety. WHO has maintained its guideline for As as the calculated amount is lower than the quantification level. Guideline values are not established for chemical concentrations that cannot be properly assessed, to be reiterated. These cases allow for the establishment of preliminary guideline values within acceptable analytical boundaries. The WHO increased the first recommendation for arsenic in potable water to 10 μg/L in 2011. Arsenic concentrations can be reduced to 5 μg/L or below by various treatment methods depending on treatment performance. Obtaining 10 μg/L utilizing traditional treatment methods such as coagulation, process control, and optimization are crucial. The World Health Organization maintained the 10-μg-per-liter (μg/L) provisional drinking water guideline in 2017, along with the analytical achievability and treatment performance criteria established in 2002. A health-based risk evaluation was not used to generate the current WHO drinking water guideline for As at 10 μg/L since it would need a lower maximum value (WHO 2002). The approach relies on routine detection limits for As in potable water and treatment technologies that have not been updated since 2011, which are methodologically outdated.

2.4 Conclusion

Finally, skin lesions, cancer, heart disease, and developmental delays are only some of the serious health concerns of arsenic toxicity. Numerous nations have accepted the threshold of 10 μg/L threshold for arsenic in potable water as a temporary guideline value established by the WHO. Preventive measures, therapeutic tools, and public awareness campaigns should accompany regulatory schemes. Portable, lab-on-a-chip devices are being developed for the detection of water contaminants. The development of affordable water quality and disease diagnostic methods, water treatment techniques, and healthcare facilities would ensure the protection of the environment and public health. Reverse osmosis, ion exchange, and adsorption are some of the established water treatment technologies that are being used successfully for removal of arsenic from polluted water supplies. However, problems still exist in areas with little resources; therefore, there has to be an on-going investigation into how to reduce costs without sacrificing quality of life, as well as extensive public health campaigns to address the underlying social, economic, and environmental causes of arsenic exposure.

Acknowledgements The authors express their heartfelt gratitude to their esteemed institutes and all individuals who have played a role, directly or indirectly, in contributing to this study.

References

Ahmad, S. A., Sayed, M. H. S. U., S. Barua, 2001. Arsenic in drinking water and pregnancy outcomes. *Environmental Health Perspectives* 109: 629–631.

Ahsan, H., Perrin, M., Rahman, A., Parvez, F., Stute, M., Zheng, Y., and J. Graziano, 2000. Associations between drinking water and urinary arsenic levels and skin lesions in Bangladesh. *Journal of occupational and environmental medicine* 42:1195–1201.

Ali, W., Rasool, A., Junaid, M., and H. Zhang, 2019. A comprehensive review on current status, mechanism, and possible sources of arsenic contamination in groundwater: a global perspective with prominence of Pakistan scenario. *Environmental geochemistry and health* 41:737–760.

Astolfi, E., Maccagno, A., García Fernández, J. C., Vaccaro, R., and R. Stimola, 1981. Relation between arsenic in drinking water and skin cancer. *Biological Trace Element Research* 3:133–143.

Chakraborti, D., Sengupta, M. K., Rahman, M. M., Ahamed, S., Chowdhury, U. K., Hossain, M. A., and Q. Quamruzzaman, 2004. Groundwater arsenic contamination and its health effects in the Ganga-Meghna-Brahmaputra plain. *Journal of environmental monitoring* 6:74N-83N.

Chakraborty, A. K., and K.C. Saha, 2013. Arsenical dermatosis from tubewell water in West Bengal. *Indian journal of medical research* 137.

Chen, C. J., Hsueh, Y. M., M.S. Lai, 1995. Increased prevalence of hypertension and long term arsenic exposure. *Hypertension* 25: 53–60.

Coronado Gonzalez, J. A., Del Razo, L. M., G. Garcia-Vargas, 2007. Inorganic arsenic exposure and type 2 diabetes mellitus in Mexico. *Environmental Research* 104:383–389.

Frisbie, S. H., and E.J. Mitchell, 2022. Arsenic in drinking water: An analysis of global drinking water regulations and recommendations for updates to protect public health. *PLoS One* 17(4): e0263505.

Guha Mazumder, D. N., Haque, R., N. Ghosh, 1998. Arsenic levels in drinking water and the prevalence of skin lesions in West Bengal, India. *International Journal of Epidemiology* 27: 871–877.

Guha Mazumder, D. N., Haque, R., N. Ghosh, 2000. Arsenic in drinking water and the prevalence of respiratory effects in West Bengal, India. *International Journal of Epidemiology* 29: 1047–1052.

Guo, X., Fujino, Y., S. Kaneko, 2001. Arsenic contamination of groundwater and prevalence of arsenical dermatosis in the Hetao plain area, Inner Mongolia, China. *Molecular and Cellular Biochemistry* 222: 137–140.

Guo, X., Liu, Z., Huang, C., and L. You, 2006. Levels of arsenic in drinking-water and cutaneous lesions in Inner Mongolia. *Journal of health, population and Nutrition* 214–220.

Hernandez-Zavala, A., Del Razo, L. M., C. Aguiler, 1998. Alteration in bilirubin excretion in individuals chronically exposed to arsenic in Mexico. *Toxicology Letters* 99: 79–84.

Hussain, A., Raveendran, V.A., Kundu, S., Samanta, T., Shunmugam, R., Pal, D., Sarma, J. Das, 2018. Mechanisms of Arsenic-Induced Toxicity with Special Emphasis on Arsenic-Binding Proteins, in: Arsenic - Analytical and Toxicological Studies. InTech.

Jha, S. K., Mishra, V. K., Damodaran, T., Sharma, D. K., and P. Kumar, 2017. Arsenic in the groundwater: Occurrence, toxicological activities, and remedies. *Journal of Environmental Science and Health Part C* 35(2): 84–103.

Liu, J., Hermon, T., Gao, X., Dixon, D., and Xiao, H., 2023. Arsenic and diabetes mellitus: a putative role for the immune system. *All Life* 16.

Milton, A. H., and M. Rahman, 2002. Respiratory effects and arsenic contaminated well water in Bangladesh. *International Journal of Environmental Health Research* 12:175–179.

Milton, A. H., Hasan, Z., A. Rahman, 2003. Noncancer effects of chronic arsenicosis in Bangladesh: Preliminary results. *Journal of Environmental Science and Health* 38: 301–305.

Milton, A. H., Smith, W., B. Rahman, 2005. Chronic arsenic exposure and adverse pregnancy outcomes in Bangladesh. *Epidemiology* 16: 82–86.

Mitchell, E., Frisbie, S., and B. Sarkar, 2011. Exposure to multiple metals from groundwater—a global crisis: geology, climate change, health effects, testing, and mitigation. *Metallomics* 3(9): 874–908.

Muzaffar, S., Khan, J., Srivastava, R., Gorbatyuk, M.S., and Athar, M., 2023. Mechanistic understanding of the toxic effects of arsenic and warfare arsenicals on human health and environment. *Cell Biology and Toxicology* 39, 85–110.

Rahman, M. M., Chowdhury, U. K., S.C. Mukherjee, 2001. Chronic arsenic toxicity in Bangladesh and West Bengal, India—a review and commentary. *Clinical Toxicology* 39: 683–700.

Rahman, A., Vahter, M., E.C. Ekstrom, 2007. Association of arsenic exposure during pregnancy with fetal loss and infant death: A cohort study from Bangladesh. *American Journal of Epidemiology* 165:1389–1396.

Santra, A., Dasgupta, J., B.K. De, 1999. Hepatic manifestations in chronic arsenic toxicity. *Indian Journal of Gastroenterology* 18:152–155.

Shirin, S., & Yadav, A. K. (2014). Physico-chemical analysis of municipal wastewater discharge in Ganga river, Haridwar district of Uttarakhand, India. Current World Environment, 9(2), 536–543. https://doi.org/10.12944/CWE.9.2.39

Smith, A. H., Lopipero, P. A., M.N. Bates, 2002. Arsenic epidemiology and drinking water standards. *Science*, 296: 2145–2146.

Tondel, M., Rahman, M., Magnuson, A., Chowdhury, I. A., Faruquee, M. H., and S.A. Ahmad, 1999. The relationship of arsenic levels in drinking water and the prevalence rate of skin lesions in Bangladesh. *Environmental health perspectives* 107:727–729.

Tsai, S. Y., Chou, H. Y., H.W. The, 2003. The effects of chronic arsenic exposure from drinking water on the neurobehavioral development in adolescence. *Neuro Toxicology* 24: 747–753.

Tseng, C. H., Chong, C. K., C.J. Chen, 1996. Dose– response relationship between peripheral vascular disease and ingested arsenic among residents in blackfoot disease endemic villages in Taiwan. *Atherosclerosis* 120: 125–133.

Tseng, C. H., Tai, T. Y., C.P. Ching, 2000. Long term arsenic exposure and incidence of non-insulin-dependent diabetes mellitus: A cohort study in arseniasis hyperendemic villages in Taiwan. *Environmental Health Perspectives* 108: 847–851.

Tucker, S. B., Lamm, S. H., Li, F. X., Wilson, R., Byrd, D. M., Lai, S., and M. Polkanov, 2001. Relationship between consumption of arsenic-contaminated well water and skin disorders in Huhhot, Inner Mongolia. *Inner Mongolia Cooperative Arsenic Project (IMCAP) Study Final Report.*

von Ehrenstein, O. S., Guha Mazumder, D. N., M.H. Smith, 2006. Pregnancy outcomes, infant mortality and arsenic in drinking water in West Bengal, India. *American Journal of Epidemiology* 163: 662–669.

Wang, C. H., Hsiao, C. K., C.L. Chen, 2007. A review of the epidemiologic literature on the role of environmental arsenic exposure and cardiovascular diseases. *Toxicology and Applied Pharmacology* 222: 315–326.

Wasserman, G. A., Liu, X., F. Parvez, 2004. Water arsenic exposure and children's intellectual function in Araihazar, Bangladesh. *Environmental Health Perspectives* 112:1329–1333.

World Health Organization, 2002. Guidelines for drinking-water quality.

World Health Organization, 2021. A global overview of national regulations and standards for drinking-water quality.

Yang, C. H., Chang, C. C., S.S. Tsai, 2003. Arsenic in drinking water and adverse pregnancy outcome in an arseniasis-endemic area in northeastern Taiwan. *Environmental Research* 91: 29–34.

Chapter 3
Mechanism and Technologies for Arsenic Bioremediation Using Natural Plant-Based Microorganism

Himanshu Tiwari and Ram Sharan Singh

Abstract Arsenic contamination leads to severe harm to the environment and public health around the globe, therefore, the development of effective and sustainable remediation technology is necessary. Conventional techniques, including physical and chemical methods for arsenic remediation, are often costly and generate enormous secondary waste, requiring complex operation and maintenance. Additionally, they can be less effective under varying environmental conditions and have a high environmental impact as a result of energy consumption and chemical residues. Therefore, this chapter comprehensively explores bioremediation technologies as a promising approach to arsenic removal, leveraging natural biological processes to detoxify contaminated environments. Bioremediation, through microbial and phytoremediation, leads to a cost-effective, environmentally friendly alternative to conventional chemical and physical remediation techniques. The chapter begins with an overview of bioremediation, highlighting the key microbial and plant-based mechanisms involved in the transformation and immobilization of arsenic. Special emphasis is placed on the role of specific potent bacteria, fungi, and hyperaccumulator plants, along with the enzymatic pathways that facilitate arsenic reduction and oxidation. A critical review of current bioremediation technologies, including 'Bioaugmentation', 'Biostimulation', and 'Constructed Wetlands', showcase their efficacy and potential practical viability. Furthermore, the environmental and biological factors influencing the efficacy of bioremediation, such as pH, temperature, redox potential, and dynamics of the microbial community, are elucidated. Future avenues of arsenic bioremediation are explored in bioprocesses that exhibit innovative solutions to mitigate research gaps and proliferate the scalability and viability of these technologies. This chapter, too, critically discusses regulatory and policy measures that are in alignment with the successful adoption of environmentally friendly bioremediation techniques. By the adoption of interdisciplinary perspectives and novel

H. Tiwari · R. S. Singh (✉)
Department of Chemical Engineering and Technology, Indian Institute of Technology (BHU), Varanasi, Uttar Pradesh, India
e-mail: rssingh.che@itbhu.ac.in

H. Tiwari
e-mail: himanshutiwari.rs.che18@itbhu.ac.in

A. K. Yadav et al. (eds.), *Arsenic Removal Technologies*,
https://doi.org/10.1007/978-3-031-91171-2_3

insights, this chapter depicts a new comprehensive insight into bioremediation technologies toward sustainable removal of arsenic, subsequentially contributing towards a healthier and safer environment.

Keywords Bioremediation · Arsenic removal · Bioaugmentation · Microbial processes · Phytoremediation · Sustainability

3.1 Introduction

Arsenic (As), a tenacious category 1 human carcinogen, is a toxic metalloid (Chauhan et al. 2024). It is found in both inorganic and organic forms. Arsine (As^{-3}), As (III), As (V), and elemental As(0) are inorganic forms, whereas arsenobetaine, dimethylarsinic acid, trimethylarsine oxide, monomethylarsonic acid, etc., are the common organic forms of As (Laha et al. 2022). It presents naturally in the Earth's crust, due to natural as well as industrial processes, especially mining have exacerbated its exposure to the environment. According to Khosravi-Darani et al. (2022), the emission of arsenic across the globe is approximately 7.8×10^7 kg/year. This release of As into the ecosystem not only contaminates aquatic life and soil but is also a severe threat to human beings across the globe. Chronic exposure to drinking water could cause various health issues especially, skin lesions, cardiovascular diseases, neurotoxicity, and various cancers. In addition, As contamination disrupts the environment, is toxic to plants and aquatic life, and steadily threatens biodiversity.

Therefore, environmentally friendly and sustainable remediation technology is a prerequisite for As removal from contaminated soil and water. Numerous conventional physicochemical techniques including adsorption, membrane filtration, and chemical precipitation are generally energy intensive and costly. Furthermore, they significantly generate secondary waste, which also needs proper disposal and treatment. Consequently, mitigation of these limitations and sustainable remediation techniques are of utmost need. Recently, bioremediation has been considered the more superior and popular technique for As remediation. It harnesses natural processes mediated by microorganisms and plants to degrade and detoxify emerging contaminants, posing a sustainable and promising alternative for As removal (Mousavi et al. 2021). Bioaccumulation, biosorption, and the generation of biosurfactants and siderophores are the microbial remediation processes of As. Bacterial strains, including *Shewanella oneidensis*, *Desulfosporosinus auripigmentum*, *Shewanella putrefaciens*, and *Desulfovibrio* strain, have the potential of As removal, using As as a terminal electron acceptor and leading to endogenous formation of As metabolites (Hussain et al. 2021).

Phytoremediation is a plant-based technology, that utilizes either naturally occurring or genetically engineered plants for the remediation of polluted environments. The phytoremediation of As is affected by the growth habit and susceptibility to the phytotoxicity associated with it and hence offers limited efficacy. The phytoremediation of As through engineering plants involves three major biotechnological

Table 3.1 A typical illustration of major sources of Arsenic release into our environment and its types (Singh et al. 2021)

Major sources	Detailed description	Arsenic types
Natural sources	The geological recede towards groundwater	As(V) and As(III)
Agricultural practices	Incorporation of pesticides, fertilizers, and herbicides with arsenic content liable for the water and soil contamination	Dimethylarsinic acid and monomethylarsonic acid (Organic arsenic compounds)
Geothermal activity	Volcanic eruption often releases geothermal fluid and affects the nearby water sources	Inorganic arsenic
Industrial activities	Mining, smelting, and coal burning mainly released arsenic into the environment	Arsenic pentaoxide, and arsenic trioxide

approaches: first, manipulation and uptake of As transport genes, production, and uptake, the As binding ligand, and finally the conversion of As to a nontoxic and volatile form.

The present chapter elucidates the mechanisms and technologies of bioremediation of As, as well as its detoxification. In addition, microbial and plant-based processes are critically reviewed and strategically explored the features of advanced bioremediation technologies. This chapter aims to assess the challenges and future scope of sustainable bioremediation of arsenic.

3.2 Background

Nowadays, As become a major global contaminant and severely affects public health globally. There are various sources of As contamination in our environment. A critical elaboration of arsenic contamination from various sources is illustrated in (Tables 3.1 and 3.2).

3.3 Conventional Arsenic Removal Technologies

Conventional methods of arsenic removal from water can be broadly classified into chemical and physical techniques (Fig. 3.1). The oxidation method oxidizes arsenite (more toxic) to arsenate and can be easily removed from the water. Chlorine, ozone, and potassium permanganate are the oxidizing agents commonly used. In coagulation-flocculation techniques, ferric chloride/aluminium sulfate is added to water, facilitating the agglomeration of arsenic particles into large particles, and removal through filtration/sedimentation (Simon et al. 2024). However, this method is effective, but the sludge generation and disposal of sludge requires careful consideration. The precipitation method uses lime/ferric salts as a chemical precipitant

Table 3.2 Major health risks associated with exposure of arsenic (Palma-Lara et al. 2020)

Health risk/related issues	Detailed description
Neurological effects	Can cause cognitive impairment, developmental delays in children, and peripheral neuropathy
Reproductive and developmental issues	Can result in adverse pregnancy outcomes, such as stillbirth, preterm birth, and low birth weight
Dermatological effects	Cause skin lesions, hyperpigmentation, and keratosis
cancer	Arsenic is a known carcinogen; prolonged exposure causes bladder, lung, skin, and liver cancers
Cardiovascular disease	Causes heart disease, stroke, and high blood pressure
Acute toxicity	Acute exposure to higher levels of arsenic might cause diarrhea, vomiting, abdominal pain, and nausea
Chronic exposure	Chronic exposure to arsenic, even at low levels, can lead to serious health problems

to separate arsenic (form of insoluble arsenic) from wastewater. It, too, generates a substantial amount of sludge and requires a proper pH adjustment during the process.

Adsorption of arsenic typically takes place through the utilization of appropriate adsorbents, such as iron oxide and activated alumina, from contaminated water. Although adsorption is highly effective but suffers from periodic regeneration/replacement and the high cost of adsorbents. Moreover, membrane-based separation techniques, such as reverse osmosis (RO) and nanofiltration (NF), are utilized for arsenic removal, and trace amounts of arsenic content can be easily removed. The main drawback of these techniques is that they generate concentrated brine (secondary waste), which suffers from disposal problems and is highly energy-intensive. The ion-exchange method used ion-exchange resins for the replacement of arsenic ions from wastewater with less harmful ions, e.g., chlorine. The periodic regeneration of resins with either strong bases or acids poses a limitation, and the presence of competing ions in water mitigates the process efficiency.

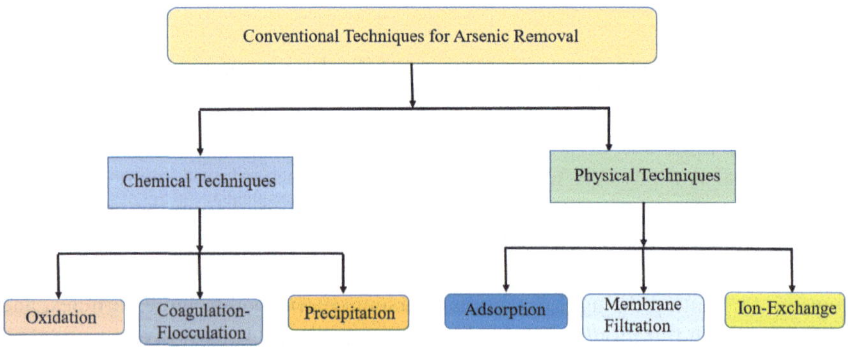

Fig. 3.1 A typical illustration of various conventional techniques used for the removal of arsenic

3.4 Fundamentals of Bioremediation

Bioremediation is the degradation, detoxification, and removal of pollutants using living organisms (microorganisms and plants). Microorganisms transform harmful pollutants into less toxic forms, using them as an energy source. However, plants absorb contaminants through their roots and transform/store them within the tissue. Plants are also known as 'hyperaccumulators'.

3.5 *In-Situ* versus *Ex-Situ* Bioremediation

The *In-situ* bioremediation method employs the remediation of pollutants at their original place. This eliminates the difficulty and need for transportation and excavation, reduces the overall cost, and makes the process economically viable (Srivastava et al. 2021). However, it prevails over the natural ecosystem through minimal disturbance to the contaminated site. Biosparging, bioventing, and Phytoremediation techniques are commonly used. In biosparging, air is injected into the groundwater, bioventing introduces air to enhance microbial activity, and phytoremediation incorporates specific plants to absorb and degrade pollutants. This process is often slower and process efficacy depends on the parameters associated with specific sites, such as soil type, plant species, and pollutant level, therefore, becomes less controllable.

In contrast, ex situ bioremediation typically involves the treatment of pollutants in a controlled environment outside of the contaminated site under precise environmental conditions that include temperature, pH, and nutrient levels (Kumar et al. 2024). Under specific environmental conditions, this technique exhibits efficient and fast outcomes. Biopiles and bioreactor are the commonly used *ex-situ* bioremediation techniques. Bioreactors are the specific systems designed for the organic matter decomposition from the contaminated soil/water using microorganisms; biopiles, where the contaminated soil is piled and allowed for aeration. The excavation and transportation of contamination not only increases the cost but also brings a significant environmental disruption.

3.6 Mechanisms of Arsenic Bioremediation

3.6.1 Microbial Processes

Microorganisms are utilizing their metabolic processes to remove arsenic from the atmosphere and convert into less toxic forms. They can either oxidize arsenic from As(III) to As(V) or reduce it from its oxidized state As(V) to its reduced state As(III), depending on environmental conditions (Kaya et al. 2024). In the reduction process, certain bacterial species including *Bacillus* and *Pseudomonas* are capable of reducing

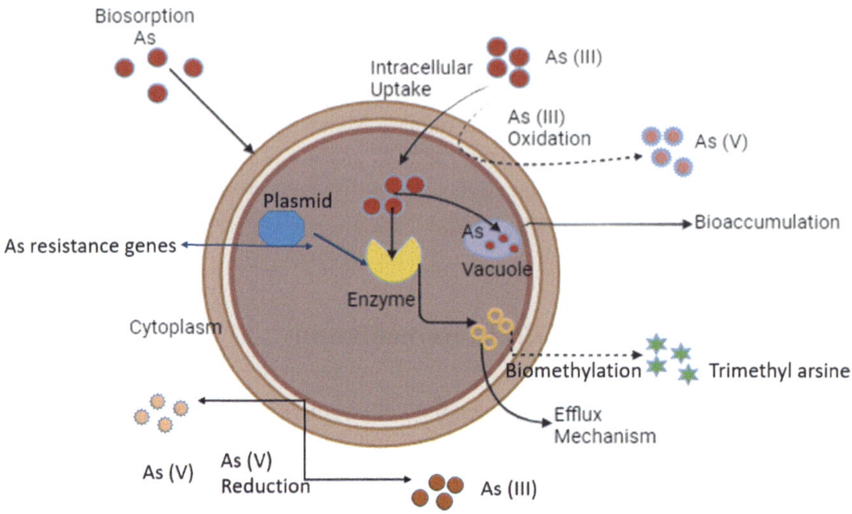

Fig. 3.2 Schematic illustration of arsenic bioremediation through microbial processes

arsenate to arsenite, while bacterial species *Leptospirillum* and *Thiobacillus* could oxidize arsenite to arsenate, respectively (Fig. 3.2).

The synergy of bioaccumulation and biotransformation can bring about a dual mechanism for the reduction of arsenic in natural systems, leveraging the ability of biological microorganisms' ability to effectively detoxify and immobilize arsenic (Qiu et al. 2024). Bioaccumulation involves the absorption and concentration of arsenic by living organisms (e.g., microbes, algae, and plants) from their surroundings to their tissues. Direct uptake, a possible bioaccumulation route, could occur from contaminated soil, water, or sediment. On the other hand, biotransformation involves the metabolic processes to convert arsenic into different chemical forms. The oxidation state of arsenic has been altered by microbes and plants either through redox reactions or by methylation, which is responsible for the successful conversion of inorganic arsenic into organic forms. These transformations are responsible for mobility and toxicity reduction, hence making it easier to stabilize or remove within the environment. Some bacteria are still capable of converting highly toxic arsenite (As(III)) into less toxic arsenate (As(V)), while some can methylate arsenic and finally produce more volatile and less toxic compounds.

3.6.2 Phytoremediation

Phytoremediation incorporates natural plant processes to extract and agglomerate arsenic from arsenic contaminated water and soils. Arsenic has been absorbed by plants through their roots and is absorbed further through their shoots and leaves.

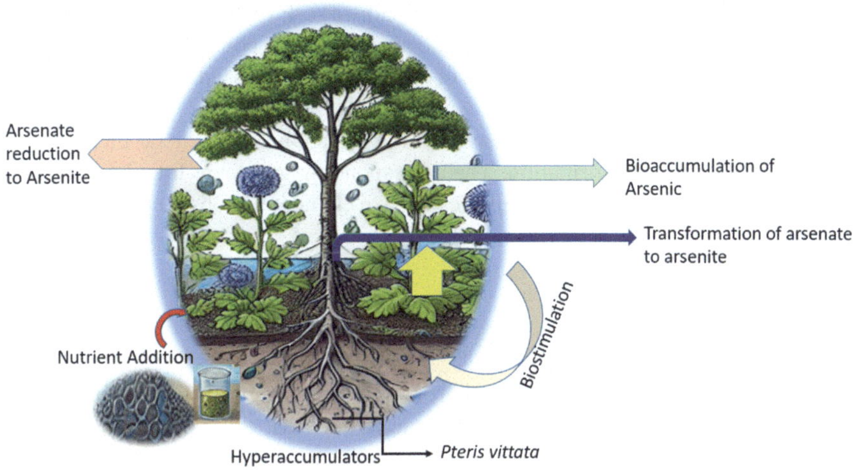

Fig. 3.3 A typical illustration of the phytoremediation mechanism of arsenic using 'Hyperaccumulators' plant Chinese brake fern (*Pteris vittata*)

Arsenic uptake occurs either actively or passively with water flow; moreover, specialized transporter proteins carry arsenic across the cell membranes (Huslina et al. 2024). Some plants are especially effective for arsenic absorption and are known as 'Hyperaccumulators'. Chinese brake ferns (*Pteris vittata*) have significant potential for high-level accumulation of arsenic in their tissues without any adverse effects (Fig. 3.3). Due to the ability of high amounts of arsenic to be stored in their biomass, these plant species are vital in phytoremediation.

The successful interactions between microbes and plants are reasonably responsible for enhancing phytoremediation. Symbiotic relationships between soil microorganisms and plants, e.g. *rhizobacteria* and *mycorrhizal* fungi, significantly proliferate plant tolerance and arsenic uptake. These microbes simply modify the chemical form of arsenic present in the soil, making it more bioavailable or less toxic to plant uptake. However, they facilitate plant growth and, subsequently, support the phytoremediation process.

3.6.3 Enzymatic Pathways

Microbes and plants implement specific enzymes to transform arsenic and significantly reduce its toxicity (Cai et al. 2023). Arsenite oxidase (Aio), Arsenate Reductase (Arr), and arsenic methyltransferase (ArsM) are the key enzymes involved in this process. Arsenite oxidase (As(V)) oxidizes arsenite (As(III)) into arsenate (As(V)), reducing its toxicity and making it easier to precipitate or immobilize. The arsenate reducase merely converts arsenate (As(V)) into arsenite (As(III)), further sequestered by plants or processed for further enzymatic pathways. Finally, arsenic methyltransferase leads to the methylation of inorganic arsenic, which converts it into less toxic organic forms. These methylated forms of arsenic can be released or volatilized from the environment.

The bioremediation process depends on the metabolic and genetic mechanisms involved in arsenic transformation. The corresponding genes are found in operons, or gene clusters, controlled by arsenic-responsive elements (Mohsin et al. 2021). For example.

3.6.3.1 Ars Operon

This group of genes includes regulatory proteins (arsR), arsenite efflux pumps (arsB), and arsenate reductase (arsC). When arsenic is exposed, this operon is activated, which helps the substance to reduce and extrude from the cell.

3.6.3.2 Oxidative Stress Response

Since arsenic causes oxidative damage, genes related to this response also play a role. Superoxide dismutase and catalase are the enzymes that mitigate this stress and promote microbial activity and survival in contaminated regions.

3.7 Bioremediation Technologies for Arsenic Removal

3.7.1 Bioaugumentation

Bioaugmentation involves the introduction of microbial consortiums or specific microorganisms into contaminated sites to enhance the degradation/transformation of contaminants such as arsenic (Chen et al. 2024). These approaches especially utilize engineered or naturally occurring microbes that can metabolize arsenic, reducing its mobility and toxicity, as well. Certain bacteria, including *Geobacter and Shewanella*, have exhibited successful introduction into arsenic-contaminated groundwater. These bacteria reduce arsenate (As(V)) to arsenite (As(III)), further adsorbed onto solid surfaces or precipitated, and reduce the arsenic concentration in the water.

However, *Pseudomonas and Thiobacillus* species are responsible for the oxidation of arsenite (As(III)) into less toxic arsenate (As(V)), facilitating the subsequent removal through conventional water treatment methods. This technique has exhibited suitability for pilot projects for drinking water treatment. Recent advancements comprise the development of microbial consortiums, especially for bioaccumulation and biosorption, with simultaneous oxidation and reduction of arsenic. These consortiums have significant potential to enhance the bioremediation of arsenic present in both soil and water.

3.7.2 Biostimulation

Biostimulation techniques utilize the intrusion of nutrients and supplements into contaminated sites, merely enhancing the activity of native microorganisms, thereby altering or breaking down of arsenic takes place (Nivetha et al. 2023). The aim is to enhance the natural bioremediation potential of these microorganisms by creating optimal bioprocess conditions for their growth and metabolism. Sensitive environmental factors such as temperature, pH, and moisture content are optimized to exhibit sustained microbial activity and ensure efficient arsenic removal.

3.8 Phytoremediation Techniques

3.8.1 Constructed Wetlands

Constructed wetlands, engineered ecosystems, are designed to effectively utilize natural processes for the remediation of pollutants such as arsenic present in water. These systems synergistically use the soils of wetland vegetation and microbial communities associated with them to exhibit efficient degradation. Plant species such as cattails and reeds play a vital role; microbial communities are supported by their roots, and so absorption and metabolization of arsenic occurs. These plants not only enhance microbial processes but also assist in stabilizing, which plays a vital role in arsenic removal.

3.8.2 Green Filters

Green filters utilize vegetated systems including riparian buffer strips or forested areas, to purify the water as it flows through the root zones. These systems take advantage of the natural ability of plants to absorb and immobilize arsenic, significantly reducing its concentration in water. Therefore, the successful adoption of these biological processes opens up a new avenue toward an environmentally friendly solution for the removal of arsenic.

3.8.3 Phytoextraction and Phytostabilization

In phytoextraction, arsenic from polluted soil is absorbed by hyperaccumulator plants, which store it in their shoots and properly leave until it is harvested and disposed of. In contrast, a process known as phytostabilization employs plants to stabilize arsenic in the soil, preventing it from migrating or seeping into groundwater or other locations. By immobilizing arsenic in the root zone, this approach reduces its environmental impact and bioavailability.

A critical review of arsenic phytoremediation is illustrated in Table 3.3.

Table 3.3 A typical illustration of the phytoremediation potential of various plant species toward Arsenic uptake

Plant species	Major experimental findings	References
Lemna valdiviana	pH: 6.3–7.0; P-PO_4: 0.0488 mmol/L; N-NO_3: 7.9 mmol/L; 1190 mg/kg As accumulation (dry weight) and 82% reduction occurs	(de Souza et al. 2019)
Marsilea quadrifolia and *Eicchornia crassipes* are effective for As removal	Bioaccumulation Factor > 1 for ten terrestrial and seven aquatic plants. *Cyanodon dactylon* and *Phyllanthus amarus exhibited a* phytoextraction potential. The highest accumulation of As was shown by *Althernanthera ficoides*	(Singh et al. 2022)
Acacia mangium	*Acacia mangium* thrived in arsenic contaminated soil having a concentration of 500 mg/kg. Phytostabilization was the leading mechanism for the uptake of arsenic	(Rosli et al. 2021)
Vallisneria natans	As(III) was oxidized to As(V) and methylated to dimethylarsinate. Arsenate was found 95.65 ± 0.10% in roots, at the end of the 14th day As(III) concentration was 3.45–6.96 mg/kg	(Li et al. 2018)
Helianthus annuus	As tolerance was 650 mg/L, 67% accumulation occurred in shoots and 55% in roots of *Helianthus annuus*	(Govarthanan et al. 2018)
Pteris vittata L	*Pteris vittata* L revealed a hyperaccumulation potential and extracted arsenic in its tissues. 44.03% mitotic depression, 16.6% relative abnormality rate, and 1.06% chromosomal abnormality frequency were less compared to those grown in unremediated soil	(Gupta et al. 2022)
Pennisetum purpureum	The highest As(V) uptake of 1549.26 ± 41.83 mg/kg, 36.36% eliminated for 5 mg As(V) kg^{-1} sand on 28 days	(Rahman et al. 2024)
Arundo donax L	The successful addition of Earth Total Support exhibited a 99.25 and 7.10% increase in As content in roots and shoots, respectively	(Zeng et al. 2023)

3.9 Factors Influencing Bioremediation Efficacy

3.9.1 Environmental Factors

Numerous environmental parameters that affect microbial activity and pollutant mineralization are critical to the success of bioremediation. The chemical composition of soil and water, pH, redox potential, and temperature are important variables.

One of the most important factors in bioremediation is the chemical composition of the soil and water. The availability of nutrients, the existence of competing pollutants, the texture of the soil, and the amount of organic matter all have a major effect on microbial activity and the bioremediation process as a whole. To promote microbial growth and activity, the necessary nutrients, including carbon, nitrogen, and phosphorus, must be available in sufficient proportions. The arsenic removal efficiency can be influenced by the presence of other contaminants, which may inhibit the bioremediation process or compete with them. The coarse-textured soils exhibit reduced surface area and limit interactions between pollutants and microorganisms, thereby reducing the efficiency of bioremediation. In contrast, organic matter-rich fine-textured soils enhance microbial activity and facilitate the adsorption of contaminants, which makes them labile for improved arsenic removal.

The microbial activity is severely affected by the pH of the soil. Most arsenic-degrading bacteria could thrive in neutral to slightly alkaline environments, optimizing their microbial processes and exhibiting effective bioremediation. However, microbial growth diminishes under extreme pH levels, reducing the bioprocess efficiency. The metabolic pathways of microbes are influenced by the redox potential. Reductive transformations are supported by anaerobic conditions (negative redox potential), whereas oxidative degradation processes are favored by aerobic conditions (positive redox potential). The availability of electron donors and acceptors in the environment controls the redox conditions and ensuing microbial activity. The usual range of ideal temperatures for bioremediation is 20–30 °C. While lower temperatures can slow microbial activity and bioremediation rates, higher temperatures can accelerate the breakdown process by enhancing microbial metabolism and enzyme activity.

3.10 Bioavailability of Arsenic

Arsenic speciation is reasonably liable for environmental mobility and bioavailability. Arsenite (As(III)) and arsenate (As(V)) are the two predominant forms and exhibit distinct behaviors. Arsenate is less toxic and more stable; however, under alkaline environments, it becomes more mobile. Arsenite is more stable, more toxic, and highly mobile, particularly under reducing conditions. The movement and transformation of these arsenic species are sensitive to soil pH, redox potential, and presence of organic matter, thus affecting their availability to microbial communities.

The mobility and bioavailability of arsenic have been significantly altered by competing ions present in water and soil. Phosphate, bicarbonate, and sulfate ions compete with arsenic for adsorption in active sites available either on microbial surfaces or soil particles. Phosphate ions are chemically similar to arsenate, and they might outcompete arsenate binding sites and enhance arsenic mobility. Furthermore, the interaction of arsenic with aluminium, iron, and manganese oxides can affect its adsorption and desorption, hence its concentration and distribution in the environment. Therefore, a better understanding of these complex interactions is vital for the development of better bioremediation techniques.

3.11 Microbial Community Dynamics

The extent of biodegradation and mineralization of pollutants such as arsenic is governed by the microbial communities, which are vital in the bioremediation process. The microbial communities comprising bacteria, archaea, and fungi interact in competitive and/or cooperative ways. This microbial diversity depicts metabolic redundancy, increasing resilience, and thereby improving bioremediation efficiency. The function and structure of microbial communities, affected by these interactions, are directly responsible for the effectiveness of bioremediation. Interactions that include competition, symbiosis, and predation could influence microbial behavior and overall bioprocess efficacy.

The dynamics of the microbial community can severely be affected by environmental stressors, including pH fluctuations, temperature changes, and exposure to toxic substances. These stressors are responsible for the reduction in microbial diversity, the suppression of the growth of sensitive species, and/or the decline of microbial activity, all of which collectively reduce the effectiveness of bioremediation. However, stress-tolerant microbial species could thrive under adverse environmental conditions, enhancing the bioremediation process. An in-depth understanding of environmental stressors on microbial populations is vital to ensure that under optimal conditions the bioremediation process could achieve a maximum degree of mineralization. A critical review of arsenic bioremediation using different microorganisms is presented in Table 3.4.

Table 3.4 A typical illustration of various microbial species for removing arsenic

Microbial species	Critical experimental findings	References
Bacillus safensis and *Lysinibacillus* sp.	Arsenic-resistant bacteria could resist 88.53 and 76.98 mM arsenite and 721.13 and 560.88 mM arsenate, respectively	(Mandal et al. 2021)
Brevibacillus sp. KUMAs2	40% As the removal occurred under aerobic conditions. *Brevibacillus* sp. KUMAs2 resisted 17 and 265 mM of As(III) and As(V), respectively	(Mallick et al. 2014)
Aneurinibacillus aneurinilyticus and *Bacillus* sp.	Bacteria could tolerate concentrations of 550 and 4500 ppm of arsenite and arsenate, respectively. 51.99 and 51.45% removal of arsenite and 50.37 and 53.29% removal of arsenate occurred	(Dey et al. 2016)
Humicola sp.	The highest biovolatilization of As 53.39 mg/kg was observed, a typical bioaccumulation range of 0.146 to 11.36 g/kg was observed	(Tripathi et al. 2020)
Shinella	The oxidation of As(III) to As(V) occurred autotrophically, and the arsenite oxidase (aioA) gene might be involved	(Nguyen et al. 2021)
Microbacterium oxydans AE038–20, $NdFeO_3$	The microbial biofilm adhering to $NdFeO_3$ showed 84.80% arsenic removal and also oxidized As(III) to As(V). The $NdFeO_3$ nanoparticles exhibited an adsorption capacity of 156 and 54.2 mg/g for As(V) and As(III), respectively	(Lascano et al. 2023)

3.12 Challenges and Limitations

Nowadays, the effective incorporation of living organisms to remove arsenic from contaminated environments severely faces various practical and technical challenges. The foremost major issue originates from the uneven distribution of arsenic contaminants in polluted areas; making uniform treatment more tedious. Environmental factors including temperature, pH, and redox conditions can vary and directly affect microbial activity and the efficacy of the bioremediation process. In addition, continuous monitoring and adjustments are required to ensure the optimum bioprocess conditions for microbial growth and arsenic removal at actual contaminated sites. Scale-up of laboratory techniques for the successful implementation at a large scale, i.e., real-world contaminated region, is another major challenge. This adequately requires the development of suitable large-scale bioreactors along with efficient systems to feed microbes or essential nutrients to contaminated regions.

The deployment of successful bioremediation techniques also requires primarily socioeconomic and regulatory considerations. Large or heavily contaminated regions often require a high cost of implementing bioremediation, therefore, leading to severe financial challenges for certain stakeholders or local communities. The adoption of genetically engineered microorganisms can influence the public perception of bioremediation toward its adoption and also raise effectiveness and safety concerns. Moreover, the regulatory frameworks for environmental remediation are very complex, vary among regions, and require adherence to standards and approval processes. Mutual understanding and collaboration among communities, researchers, and policymakers will surpass these existing challenges and develop a smooth framework that ensures that bioremediation strategies are also socially acceptable and effective, too.

3.13 Future Directions and Research Needs

Recent advances in Bioremediation could transform arsenic contamination removal strategies through the utilization of innovative microbial species, advanced bioreactors, and bioaugmentation techniques to improve efficiency. The successful integration of bioremediation with cutting-edge technologies laid the foundation stone of Hybrid Techniques that potentially eradicate the arsenic contaminant. 'Nanobiotechnology', 'Bioinformatics', and 'Metagenomics' are recently emerging and are leading to new avenues towards the existing bioremediation technology. Nanomaterials can proliferate microbial activity and enhance arsenic adsorption, while bioinformatics plays a vital role in the better understanding of microbial metabolic pathways and their regulation. Metagenomics is responsible for the identification and optimization of the microbial communities involved in arsenic transformation.

A synergistic multidisciplinary approach that integrates expertise from various domains of engineering, environmental science, microbiology, and chemistry is crucial for effective bioremediation. This interdisciplinary collaboration will lead to a comprehensive solution to the challenges posed by arsenic contamination. This strategy also needs consideration of socioeconomic and policy factors to ensure long-term adoption. Research gaps must be addressed to modify arsenic removal technologies, including the development of scalable and cost-effective methods, understanding how environmental factors influence microbial communities and promising the long-term stability of bioremediation. Furthermore, advanced research is also required on microbial arsenic resistance, microbial interactions in bioremediation, detoxification mechanisms, and the major environmental impacts.

Synthetic biology and genetic engineering symbiotically become effective bioremediation tools. Through the modification of genetic pathways, researchers can develop microorganisms that have mammoth resistance to arsenic along with improved capacity for adsorption and transformation. It also creates microbial strains that are exceptionally more resilient to harsh environmental conditions. However, strict policies and regulations are safer and need effective bioremediation deployment, especially for the adoption of genetically modified microorganisms. These clear and strict guidelines must address public concerns and encourage research and development for the removal of arsenic.

Funding, support grants, and subsidies are predominantly requisite for those policies that overwhelmingly encourage bioremediation technology and also bring its widespread adoption. Highly arsenic-contaminated regions also require a flexible regulatory framework that facilitates the smooth implementation and approval of bioremediation projects. Successful cooperation among industries, academia, and industries will foster a supportive environment towards the development of bioremediation techniques. It is beneficial for policy makers to raise public awareness, raise the benefits of bioremediation approaches, and prioritize funding for bioremediation research. Promoting international collaboration and building alliances might strengthen bioremediation techniques for the effective removal of arsenic.

3.14 Conclusions

This chapter of the book critically explores the implementation of bioremediation for sustainable, environmentally friendly, and economically feasible arsenic removal, highlighting the main drawbacks of conventional methods. It elucidates the fundamental aspects of bioremediation and examines key remediation techniques such as ‘Bioaugmentation’ and ‘Phytoremediation’. This study also shows how arsenic bioavailability, microbial dynamics, and environmental factors affect the extent of bioremediation. The future scope of this field may lead to the adoption of new technologies and interdisciplinary approaches that integrate numerous scientific domains. Recent advances in synthetic biology and genetic engineering have gigantic potential to proliferate microbial and plant ability toward arsenic removal. The overcoming of current research gaps and regulatory challenges will be crucial in the practical adoption and advancement of these technologies. In addition, successful collaboration among scientists, local communities, and policymakers is essential for the development and proliferation of sustainable bioremediation solutions. Policies that improve research and encourage the use of efficient environmentally friendly bioremediation techniques will reduce arsenic contamination and protect the environment.

Acknowledgements The author deeply acknowledges the research facility provided by the Department of Chemical Engineering & Technology, Indian Institute of Technology (BHU) Varanasi, India.

References

Cai, X., Xue, C., Owens, G., Chen, Z., 2023. Removal of As(III) using a microorganism sustained secrete laccase-straw oxidation system. J. Hazard. Mater. 448. https://doi.org/10.1016/j.jhazmat.2023.130967

Chauhan, R., Awasthi, S., Tiwari, P., Upadhyay, M.K., Srivastava, S., Dwivedi, S., Dhankher, O.P., Tripathi, R.D., 2024. Biotechnological strategies for remediation of arsenic-contaminated soils to improve soil health and sustainable agriculture. Soil Environ. Heal. 2, 100061. https://doi.org/10.1016/j.seh.2024.100061

Chen, P., Liu, Y., Sun, G.X., 2024. Evaluation of water management on arsenic methylation and volatilization in arsenic-contaminated soils strengthened by bioaugmentation and biostimulation. J. Environ. Sci. (China) 137, 515–526. https://doi.org/10.1016/j.jes.2023.02.023

de Souza, T.D., Borges, A.C., Braga, A.F., Veloso, R.W., Teixeira de Matos, A., 2019. Phytoremediation of arsenic-contaminated water by *Lemna Valdiviana*: An optimization study. Chemosphere 234, 402–408. https://doi.org/10.1016/j.chemosphere.2019.06.004

Dey, U., Chatterjee, S., Mondal, N.K., 2016. Isolation and characterization of arsenic-resistant bacteria and possible application in bioremediation. Biotechnol. Reports 10, 1–7. https://doi.org/10.1016/j.btre.2016.02.002

Govarthanan, M., Mythili, R., Selvankumar, T., Kamala-Kannan, S., Kim, H., 2018. Myco-phytoremediation of arsenic- and lead-contaminated soils by *Helianthus annuus* and wood rot fungi, *Trichoderma* sp. isolated from decayed wood. Ecotoxicol. Environ. Saf. 151, 279–284. https://doi.org/10.1016/j.ecoenv.2018.01.020

Gupta, K., Srivastava, S., Saxena, G., Kumar, A., 2022. Application of *Pteris vittata* L. for phytoremediation of arsenic and biomonitoring of the process through cyto-genetic biomarkers of *Trigonella foenum-graecum L.* Physiol. Mol. Biol. Plants 28, 91–106. https://doi.org/10.1007/s12298-022-01124-4

Huslina, F., Khudur, L.S., Shah, K., Surapaneni, A., Netherway, P., Ball, A.S., 2024. Mine Site Restoration: The Phytoremediation of Arsenic-Contaminated Soils. Environ. - MDPI 11, 1–22. https://doi.org/10.3390/environments11050099

Hussain, M.M., Bibi, I., Niazi, N.K., Shahid, M., Iqbal, J., Shakoor, M.B., Ahmad, A., Shah, N.S., Bhattacharya, P., Mao, K., Bundschuh, J., Ok, Y.S., Zhang, H., 2021. Arsenic biogeochemical cycling in paddy soil-rice system: Interaction with various factors, amendments and mineral nutrients. Sci. Total Environ. 773, 145040. https://doi.org/10.1016/j.scitotenv.2021.145040

Kaya, C., Uğurlar, F., Ashraf, M., Hou, D., Kirkham, M.B., Bolan, N., 2024. Microbial consortia-mediated arsenic bioremediation in agricultural soils: Current status, challenges, and solutions. Sci. Total Environ. 917. https://doi.org/10.1016/j.scitotenv.2024.170297

Khosravi-Darani, K., Rehman, Y., Water, I.K.-, 2022, undefined, n.d. Arsenic exposure via contaminated water and food sources. mdpi.comK Khosravi-Darani, Y Rehman, IA Katsoyiannis, E Kokkin. AI ZouboulisWater, 2022•mdpi.com.

Kumar, V., Thakur, M., Seth, C.S., 2024. In situ remediation techniques for removal of arsenic in the environment. Curr. Opin. Environ. Sci. Heal. 38. https://doi.org/10.1016/j.coesh.2024.100538

Laha, A., Sengupta, S., Bhattacharya, P., Mandal, J., Somnath Bhattacharyya, ·, Kallol Bhattacharyya, ·, 2022. Recent advances in the bioremediation of arsenic-contaminated soils: a mini review. World J. Microbiol. Biotechnol. 38, 189. https://doi.org/10.1007/s11274-022-03375-5

Lascano, G.A., Spuches, C.F., Gómez, M.I., Ferrero, M.A., Sales, A., Romero, C.M., Navarro, M.C., 2023. Bioremediation of arsenic using magnetic $NdFeO_3$ nanoparticles functionalized with microbial biofilm. J. Environ. Chem. Eng. 11. https://doi.org/10.1016/j.jece.2023.109532

Li, B., Gu, B., Yang, Z., Zhang, T., 2018. The role of submerged macrophytes in phytoremediation of arsenic from contaminated water: A case study on *Vallisneria natans* (Lour.) Hara. Ecotoxicol. Environ. Saf. 165, 224–231. https://doi.org/10.1016/j.ecoenv.2018.09.023

Mallick, I., Hossain, S.T., Sinha, S., Mukherjee, S.K., 2014. *Brevibacillus* sp. KUMAs2, a bacterial isolate for possible bioremediation of arsenic in rhizosphere. Ecotoxicol. Environ. Saf. 107, 236–244. https://doi.org/10.1016/j.ecoenv.2014.06.007

Mandal, D., Sonar, R., Saha, I., Ahmed, S., Basu, A., 2021. Isolation and identification of arsenic resistant bacteria: a tool for bioremediation of arsenic toxicity. Int. J. Environ. Sci. Technol. 9883–9900. https://doi.org/10.1007/s13762-021-03673-9

Mohsin, H., Shafique, M., Rehman, Y., 2021. Genes and Biochemical Pathways Involved in Microbial Transformation of Arsenic. Arsen. Toxic. Challenges Solut. 391–413. https://doi.org/10.1007/978-981-33-6068-6_15

Mousavi, S.M., Hashemi, S.A., Iman Moezzi, S.M., Ravan, N., Gholami, A., Lai, C.W., Chiang, W.H., Omidifar, N., Yousefi, K., Behbudi, G., 2021. Recent Advances in Enzymes for the Bioremediation of Pollutants. Biochem. Res. Int. 2021.https://doi.org/10.1155/2021/5599204

Nguyen, V.K., Nguyen, D.D., Ha, M.G., Kang, H.Y., 2021. Potential of versatile bacteria isolated from activated sludge for the bioremediation of arsenic and antimony. J. Water Process Eng. 39. https://doi.org/10.1016/j.jwpe.2020.101890

Nivetha, N., Srivarshine, B., Sowmya, B., Rajendiran, M., Saravanan, P., Rajeshkannan, R., Rajasimman, M., Pham, T.H.T., Shanmugam, V.K., Dragoi, E.N., 2023. A comprehensive review on bio-stimulation and bio-enhancement towards remediation of heavy metals degeneration. Chemosphere 312. https://doi.org/10.1016/j.chemosphere.2022.137099

Palma-Lara, I., Martínez-Castillo, M., Quintana-Pérez, J.C., Arellano-Mendoza, M.G., Tamay-Cach, F., Valenzuela-Limón, O.L., García-Montalvo, E.A., Hernández-Zavala, A., 2020. Arsenic exposure: A public health problem leading to several cancers. Regul. Toxicol. Pharmacol. 110, 104539. https://doi.org/10.1016/j.yrtph.2019.104539

Qiu, D., Yu, Z., Zhang, X., Wen, C., Yan, C., 2024. Influence of extracellular polymeric substances on arsenic bioaccumulation and biotransformation in biofilms. Chemosphere 349. https://doi.org/10.1016/j.chemosphere.2023.140798

Rahman, M.E., Uddin, M.K., Shamsuzzaman, S.M., Mahmud, K., Shukor, M.Y.A., Ghani, S.S.A., Nabayi, A., Sadeq, B.M., Chompa, S.S., Akter, A., Halmi, M.I.E. Bin, 2024. Potential use of *Pennisetum purpureum* for phytoremediation of arsenic in treatment sand: A phytotoxicity study. Biocatal. Agric. Biotechnol. 60, 103300. https://doi.org/10.1016/j.bcab.2024.103300

Rosli, R.A., Harumain, Z.A.S., Zulkalam, M.F., Hamid, A.A.A., Sharif, M.F., Mohamad, M.A.N., Noh, A.L., Shahari, R., 2021. Phytoremediation of Arsenic in Mine Wastes by *Acacia mangium*. Remediat. J. 31, 49–59. https://doi.org/10.1002/rem.21688

Simon, M., Joshi, H., Yadav, A. K., & Hopke, P. K. (2024). Simultaneous nitrogen and phosphorus removal from polluted water using a novel bacterial consortium (BACON-38): Promising application for wastewater remediation. Journal of Water Process Engineering, 66, 106044. https://doi.org/10.1016/j.jwpe.2024.106044

Singh, P., Borthakur, A., Singh, R., Bhadouria, R., Singh, V.K., Devi, P., 2021. A critical review on the research trends and emerging technologies for arsenic decontamination from water. Groundw. Sustain. Dev. 14, 100607. https://doi.org/10.1016/j.gsd.2021.100607

Singh, S., Karwadiya, J., Srivastava, S., Patra, P.K., Venugopalan, V.P., 2022. Potential of indigenous plant species for phytoremediation of arsenic contaminated water and soil. Ecol. Eng. 175. https://doi.org/10.1016/j.ecoleng.2021.106476

Srivastava, A.K., Singh, R.K., Singh, D., 2021. Microbe-based bioreactor system for bioremediation of organic contaminants: present and future perspective. Microbe Mediat. Remediat. Environ. Contam. 241–253. https://doi.org/10.1016/B978-0-12-821199-1.00020-1

Tripathi, P., Khare, P., Barnawal, D., Shanker, K., Srivastava, P.K., Tripathi, R.D., Kalra, A., 2020. Bioremediation of arsenic by soil methylating fungi: Role of humicola sp. strain 2WS1 in amelioration of arsenic phytotoxicity in *Bacopa monnieri* L. Sci. Total Environ. 716. https://doi.org/10.1016/j.scitotenv.2020.136758

Zeng, D., Pu, G., Chen, X., Wang, K., Mo, L., Liao, J., 2023. Phytoremediation of arsenic contaminated soils by *Arundo donax* L. enhanced with Earth Total Support (ETS). Sustain. Chem. Environ. 3. https://doi.org/10.1016/j.scenv.2023.100036

Chapter 4
Utilizing Natural Processes for Sustainable Arsenic Removal Through Bioremediation Technology

Navaneeth K. Nampoothiri, Ramireddy Venkata Sai Reddy, and Sourav Maity

Abstract Given its high toxicity and widespread presence in the Earth's crust, arsenic, a metalloid, poses a significant threat to human health, mainly when consumed by contaminated water. In this chapter, we extensively examine the many oxidation states of arsenic, focusing specifically on the more dangerous forms, notably arsenite (As(III)) and arsenate (As(V)), along with the corresponding toxicological consequences. This paper reveals the limitations of conventional methods for removing arsenic. These processes often lead to the generation of hazardous by-products and are reported to be ineffective in natural environments. On the contrary, biological treatment methods such as bioremediation, biosorption, bioaccumulation, and phytovolatilization have emerged as viable alternatives that are efficient and ecologically sustainable. Microbial involvement is essential in these processes, as well as biotransformation and biomineralization, to mitigate arsenic toxicity and help remove arsenic from contaminated areas. This chapter explores the genetic foundation of arsenic biotransformation, emphasizing the crucial role of specific genes and microbial interactions in detoxification. Furthermore, it emphasizes the possibility of using these inherent processes to create novel and ecologically sound interventions to deal with arsenic pollution in different environmental settings.

Keywords Arsenic detoxification · Arsenic toxicity · Microorganisms · Biosorption · Biotransformation · Bioaccumulation

N. K. Nampoothiri
Department of Biochemical Engineering and Biotechnology, Indian Institute of Technology Delhi, Hauz Khas, New Delhi, Delhi, India

R. V. S. Reddy
Centre for Biomedical Engineering, Indian Institute of Technology Delhi, Hauz Khas, New Delhi, Delhi, India

S. Maity (✉)
Centre for the Environment, Indian Institute of Technology Guwahati, Guwahati, Assam, India
e-mail: sourav.maity.92@gmail.com

A. K. Yadav et al. (eds.), *Arsenic Removal Technologies*,
https://doi.org/10.1007/978-3-031-91171-2_4

4.1 Introduction

Arsenic (As), a highly poisonous pollutant, occurs naturally in the Earth's crust, with an average concentration of 1.8 parts per million (ppm). It belongs to the 15th group of elements in the periodic table, specifically in the 4th period (Nurchi et al. 2020; Sanyal et al. 2020). Arsenic is classified as a group I carcinogen and poses a significant public health problem, primarily due to consuming contaminated water (Maity et al. 2024). While acute arsenic poisoning can present with anemia, diarrhea, and gastrointestinal upset, acute arsenic poisoning can cause skin lesions, neurological disorders, and various cancers (Bhat et al. 2024). As of 2020, arsenic is the most toxic chemical listed by the US Agency for Toxic Substances and Disease Registry (ATSDR) (Maity, Biswas, and Sarkar 2020). The World Health Organization (WHO) has set a recommendation that states that the highest allowable amount of arsenic in drinking water should not exceed 10 μg per liter (μg/L) (Maity et al. 2025; Dokania et al. 2024). Arsenic levels in soil have been shown to fluctuate, with typical background values ranging from 0.1 to 10 mg per kilogram (mg/kg) and an average concentration of approximately 5 mg/kg. However, due to global conditions, there may be elevated concentrations of arsenic in certain areas. On average, the levels range from 2 to 20 mg/kg, but in exceptional cases, they can reach up to 40 mg/kg. On the other hand, soil polluted by human activities such as mining, smelting operations, or using pesticides containing arsenic can have significantly elevated levels of arsenic, ranging from 5 to 3000 mg/kg (Jimenez et al. 2023) (Table 4.1). Arsenic can exist in many oxidation states; the most common forms are aquatic and highly carcinogenic arsenite (As(III)) and arsenate (As(V)). The other two oxidation states are nonaquatic arsenic (As(0)) and Arsine (AsH3), which are rare in aquatic systems. The redox potential and pH primarily influence the distribution of inorganic arsenic species. In oxidizing surface waters, As(V) is the predominant form, while in reducing groundwater, As(III) is more stable. As(III) interacts less with solid surfaces than As(V), making it more difficult to remove using conventional treatment methods (Maity et al. 2021). Arsenic compounds exhibit varying levels of toxicity, with arsine gas and organic trivalent forms being the most toxic, followed by inorganic arsenite, inorganic arsenate, organic pentavalent compounds, and finally, elemental arsenic and arsonium compounds (Irshad et al. 2021).

However, most traditional methods for purifying arsenic have practical limitations, including generating dangerous sludge, ineffectiveness in natural environments, expensive operation and maintenance, and operational challenges (Maity et al. 2022; Yadav et al. 2014). Biological treatment has received significant attention because of its advantages over standard physicochemical treatments. Biological treatment is the employment of microorganisms to reduce, oxidize, or eliminate toxins in groundwater rather than relying on chemical reagents (Dokania et al. 2024). The promotion of repair depends on alterations in the oxidation–reduction potential, commonly referred to as the redox potential. Microorganisms can facilitate the removal of certain metals from groundwater sources through a process called bioreduction. Although the effectiveness of biological processes in removing

Table 4.1 Total number of nations impacted by groundwater arsenic pollution on each continent ($\geq$10 μg/L) and the corresponding percentage of affected regions and people (Bhat et al. 2024)

Continent	Proportion of globally affected areas	Countries affected	Proportion of total global affected population
Asia	64%	31	94%
Oceania	2%	4	0.01%
Africa	9%	20	3.7%
South America	14%	9	1.6%
North America	10%	11	0.6%
Europe	1%	31	0.2%

arsenic is still being studied, bacterial activity is a crucial catalyst in various arsenic removal methods (Hubadillah et al. 2020). Several recent studies have shown that the removal of arsenic and iron can be achieved simultaneously by using biological filtration under optimal pH, temperature, and oxygen conditions. The primary factor that significantly influences the success of arsenic removal is the initial concentration of iron. Higher iron concentrations yield more than 90% removal effectiveness, while lower concentrations only reach approximately 40%. Supplementing systems with low iron contents with ferrous sulfate is advisable to remove the arsenic that has been accomplished completely. An essential process involves the immobilization of As^{+3} in iron oxides facilitated by bacterial action (Roy et al. 2023).

4.2 Bioremediation of Arsenic

Arsenic, a metalloid with known hazardous properties, is primarily present in two oxidation states: arsenite (As^{3+}) and arsenate (As^{5+}) (Maity et al. 2021). Arsenite is more toxic than arsenate because of its ability to disrupt protein function by binding to sulfhydryl groups. This leads to a decrease in cysteine residues and hinders the proper folding of proteins. Its method of action inhibits phosphate transporters. Although arsenate also poses potential hazards. The conversion of arsenate to arsenite, leading to increased toxicity, is the primary cause of concern regarding arsenate (Dokania et al. 2024). Studies have revealed that arsenite has been found in settings that resemble primordial habitats, such as thermal vents. This discovery implies that primitive bacteria may have used these ions for bioenergetic processes, uncovering an ancient connection between arsenic and life (William and Magpantay 2023).

Prokaryotic microorganisms have evolved various resistance mechanisms to mitigate the detrimental consequences, illustrated in Fig. 4.1. By undergoing redox reactions, they can alter the chemical properties of arsenic, converting As^{3+} to As^{5+} during oxidation and reversing the process during reduction (Roy et al. 2024). Moreover, these microorganisms can enhance the solubility of toxic metals, thereby facilitating their removal from polluted environments. In addition, they can immobilize

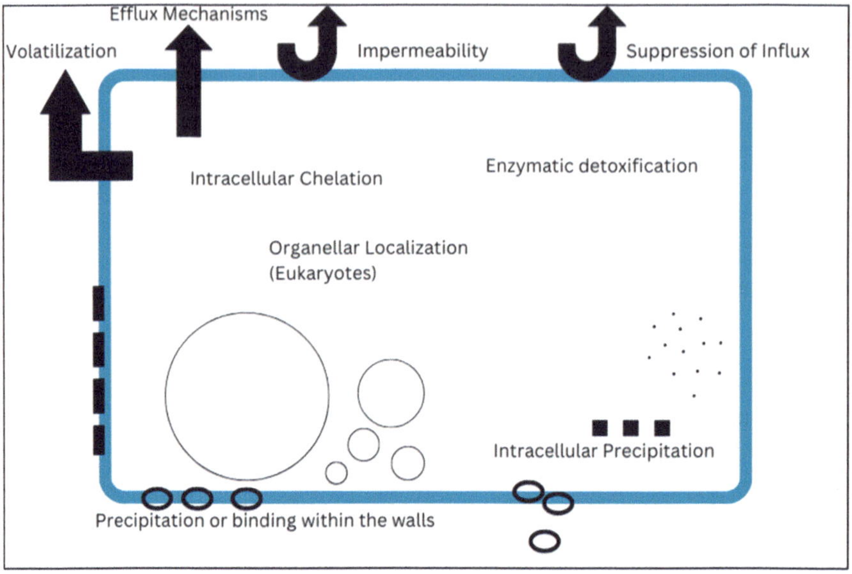

Fig. 4.1 Various microbial resistance mechanisms to mitigate arsenic's detrimental consequences

metal ions by causing them to precipitate out of the soil solution. This reduces the availability of arsenic and minimizes its adverse effects (Rahman and Singh 2020). There are multiple mechanisms involved in the microbial bioremediation of arsenic-contaminated wastewater, such as biotransformation, biomineralization, precipitation, bioleaching, biosurfactant production, biomethylation, volatilization, biosorption, bioaccumulation, and microbe-assisted phytoremediation (Gadd 2010; Simon et al. 2024) as illustrated in Fig. 4.2.

4.2.1 Biosorption

Biosorption is how biological substances, particularly bacteria, remove heavy metals and other compounds from their immediate surroundings by adsorbing in a solution (Dokania et al. 2024). Biosorption may involve chemical or physical bonds with specific functional groups. Microprecipitation, electrostatic attraction, covalent bonding, complexation, ion exchange, and van der Waals forces are critical mechanisms that contribute to the effectiveness of biosorption. Biosorption processes may be primarily categorized as (i) metabolism-dependent and (ii) non-metabolism-dependent, as illustrated in Fig. 4.3.

Biosorption is the phenomenon in which metal ions interact with functional groups on the surface of microbial cells. This engagement may occur through several mechanisms. The occurrence of biosorption on the cell surface is facilitated by

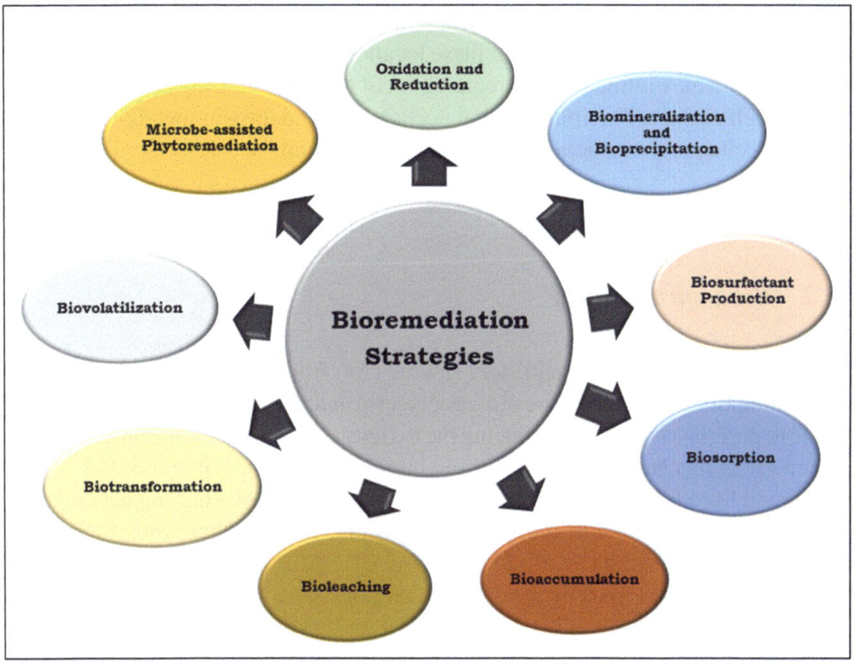

Fig. 4.2 Different Techniques of Bioremediation

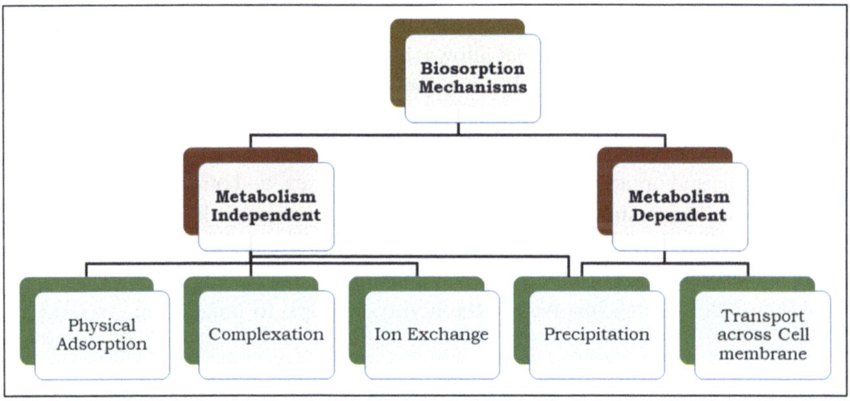

Fig. 4.3 Biosorption processes are categorized according to their dependence on the microbial cells' metabolic activity

ion exchange, complexation, and physical adsorption. These activities do not need metabolic activity. Cell walls of microbial biomass, composed of polysaccharides, proteins, and lipids, provide many sites for sorption, ion exchange, and covalent binding. The carboxyl, hydroxyl, sulfhydryl, amino, and phosphate groups only illustrate the functional groups present at these locations (Maity et al. 2023). The cellular metabolic state may influence the precipitation of metals, occurring either in solution or on the cell surface. Precipitation that occurs independently of metabolism results from direct chemical interactions between metal ions and the cell surface (Limo et al. 2018).

4.2.2 Bioaccumulation

Microorganisms can capture and store arsenic (As) from their environment through bioaccumulation. This procedure utilizes glycerol and phosphate transporters to bind arsenic to proteins or peptides, reducing the toxicity of the arsenic substance. Bacteria use numerous mechanisms, including ion pumps, carrier-mediated transport, endocytosis, and lipid infiltration, to promote bioaccumulation. Arsenic can adhere to the surfaces of microorganisms and then be transported within their cells by specialized transporters (Preetha et al. 2023; Yin et al. 2022) (Table 4.2). Arsenite (As (III)) can be taken up by cells via aquaglyceroporins, such as GlpF in bacteria and Fps1p in yeast, where it bonds with sulfhydryl groups in proteins. Phosphate transporters may absorb Arsenate (As(V)) due to its structural similarities with phosphate. *Escherichia coli* utilizes the pentavalent phosphate transporters Pit and Pst to enhance the transfer of arsenate. Pit serves as the primary protein responsible for absorption in this process. The transport mechanism is driven by a proton motive force, which enables the exchange of internal phosphate with extracellular phosphate or arsenate. Pst is an ABC-type transporter that allows the entry of both phosphate and arsenate. However, it is less efficient than Pit in transporting both substances. When arsenate is present, Pst also inhibits phosphate transport (Maity, Bose, et al. 2024).

An aquaglyceroporin, or GlpF, is a channel protein primarily responsible for transporting uncharged organic solutes such as glycerol, urea, and other small molecules. Recent studies have shown that GlpF also plays a crucial role in facilitating the transportation of arsenite (As (III)) across bacterial cell membranes. Understanding arsenic absorption and toxicity in microorganisms is of significant significance, and one of the mechanisms involved is the ability of GlpF to translocate As(III) (Yin et al. 2022).

4.2.3 Biotransformation

Arsenic may exist in four primary oxidation states: -3 (as arsine), 0 (as elemental arsenic), +3 (as arsenite), and +5 (as arsenate). Arsenite and arsenate are the most common forms of arsenic in the environment. Arsenite poses a far greater risk than arsenate due to its ability to interact with the thiol groups of proteins and enzymes,

Table 4.2 Microbial strains with tolerance to arsenic (Preetha et al. 2023)

No	Strain isolate	Tolerance
1	*Brevundimonas aurantiaca* PFAB1	Up to 90 mM arsenite Up to 310 mM arsenate
2	*Aeromonas*	Up to 100 mM arsenate Up to 20 mM arsenite
3	*Microbacterium*	Up to 69.2 mM arsenite
4	*Alicyclobacillus mali* FL 18	Up to 41 mM arsenite
5	*Bacillus* sp. *KM02*, *Aneurinibacillus aneurinilyticus*	Up to 4500 mg/l arsenate Up to 550 mg/l arsenite
6	*Bacillus* sp. *KL6*	Up to 90 mg/l arsenic
7	*Pseudomonas* sp. QNC1	Up to 350 mM arsenate
8	*Klebsiella pneumonia* sp.	Up to 28 mM of arsenic
9	*Brevibacterium linens* strain AE038-8	Up to 1 M of arsenate Up to 75 mM of arsenite
10	*Lysinibacillus* sp., *Bacillus safensis*	Up to 88.53 mM arsenite Up to 721.13 mM arsenate
11	*Acromobacter xylosoxidans*	Up to 15 mM As(III)

suppressing their activity. Furthermore, studies have shown that arsenic has mutagenic and genotoxic properties in humans (Biswas et al. 2023). Microorganisms have evolved several mechanisms, including arsenite oxidation, arsenate reduction, and arsenite methylation, to mitigate the toxicity of arsenic. The regulation of these systems is governed by specific genes (aio, arr, arsC, and arsM) located in chromosomal or plasmid DNA. Microbes demonstrate their ability to adapt to arsenic-contaminated environments by expediting the conversion of the highly toxic form of arsenic, As (III), to the less harmful form, As(V), by oxidation. In contrast, they can also facilitate reducing As(V) back to As(III) (Mujawar et al. 2019; Rahman and Singh 2020).

4.2.3.1 Genetic Basis of Arsenic Biotransformation

The aio gene, which encodes the arsenite oxidase enzyme, is very prevalent in bacteria engaged in arsenic biotransformation. This gene comprises a maximum of seven subunits: aioR, aioS, aioX, aioB, aioA, aioC, and aioD (Maity, Bose, et al. 2024). Subunits aioA and aioB are the main catalysts that facilitate the conversion of As (III) to the less dangerous form of As (V) by oxidation. Several bacteria, such as Alcaligenes, Achromobacter, Pseudomonas, Variovorax, Agrobacterium, Azoarcus, Sinorhizobium, and Azospirillum, have shown considerable capability in converting arsenic (III) to a more oxidized form by activating the aio gene (Rahman and Singh 2020). The As detoxification mechanism by the Aiox operon is illustrated in Fig. 4.4.

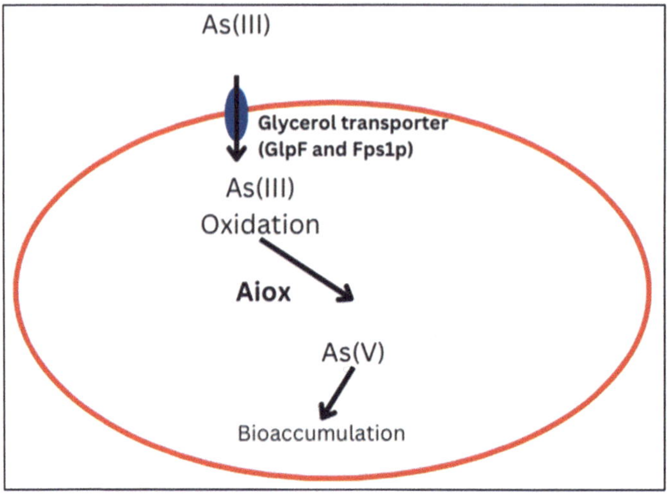

Fig. 4.4 As-detoxification mechanism of As-tolerant bacterium by the Aiox operon

The arx system is species-specific since it has only been discovered in three *gammaproteobacteria of* the *Ectothiorhodospiraceae* family. On the contrary, arsenite oxidases belonging to the ArxA category are present in a broader array of environments, indicating that arsenotrophy significantly contributes to the biogeochemical cycling of arsenic. Due to this, there are opportunities for further investigation into how these oxidases contribute to the cycling and alteration of arsenic in various ecological environments (Lis et al. 2013).

The arr genes enable dissimilatory reduction of As (V) by producing a membrane arsenate reductase, facilitating periplasmic anaerobic respiration of As(V). This metabolic process utilizes a diverse range of electron sources. Currently, there is a lack of redox data. Therefore, it is not possible to determine the values of Arr or Arx (Lis et al. 2013).

Arsenate (As(V)) and arsenite (As(III)) are taken up by cells with the assistance of several enzymes and transporter proteins, such as phosphate (Pi) transporters and aquaglyceroporins. On the contrary, the ArsA and ArsB (ArsAB) proteins create an ATP-dependent efflux pump that actively removes As(III) from the cell. The ars operon controls the manifestation of various processes associated with arsenic. It contains genes for a transacting repressor, an arsenite permease pump, and an arsenate reductase enzyme. Arsenate reductase (ArsC) employs glutathione (GSH) as a cofactor to facilitate the reduction of As(V) to the more harmful As(III) form. This form is then eliminated from the cell by the ArsAB efflux pump (Pandey et al. 2021). The mechanism of action of the Ars operon is illustrated in Fig. 4.5.

Biomethylation converts inorganic and organic arsenic forms to volatile organic methylated forms, including mono, di, and trimethylarsine. This change occurs through the activity of microorganisms. Arsenic methylation, a process seen in methanogenic bacteria, can serve as a detoxification mechanism due to its association

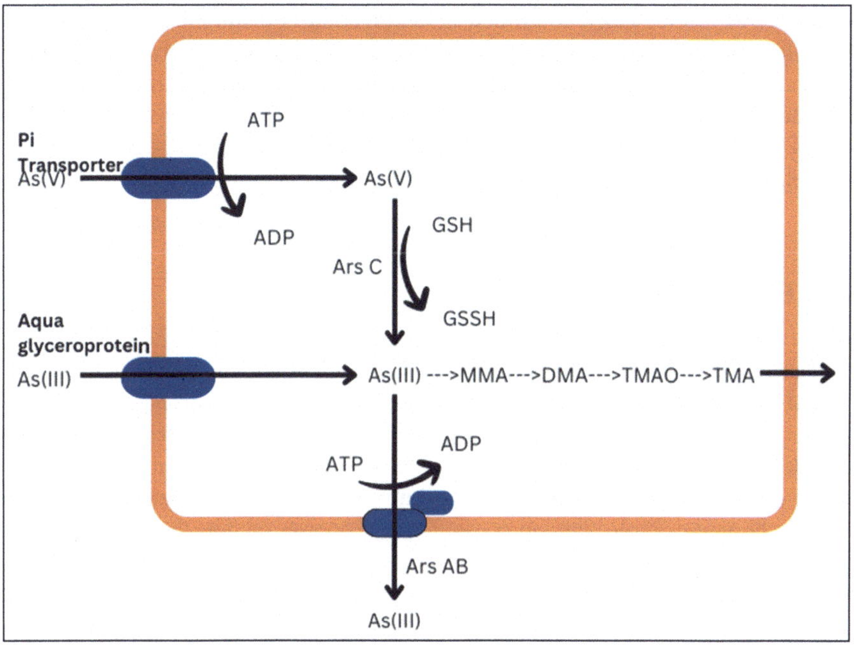

Fig. 4.5 As-detoxification mechanism of As-tolerant bacterium by the Ars operon

with methane production. An essential element of biomethylation is the production of volatile and non-volatile methylated arsenic molecules. The volatile species consist of the general structure $(CH_3)nAsH_3 - n$, where n represents the number of methyl groups, which may be 1, 2, or 3. Furthermore, other non-volatile compounds are also generated alongside methyl arsenate and dimethylarginine. The environmental biogeochemical cycle of this metalloid is significantly impacted by the extensive transformation of arsenic in microorganisms (Rahman and Singh 2020).

4.3 Biomineralization and Bioprecipitation

There are two distinct types of biomineralization: biologically induced mineralization (BIM) and biologically controlled mineralization (BCM). Biomineralization refers to the biological process by which living organisms produce minerals. Biologically induced mineralization occurs when an organism modifies its surroundings to facilitate the chemical precipitation of minerals outside its cell through mineralization. This process involves metabolic processes, which include altering the levels of pH and pCO_2 and the composition of released products. These processes indirectly influence the formation of mineral phases due to their impact along with bioprecipitation. Bioprecipitation approaches using inorganic iron and Sulfur may effectively eliminate As. Biologically controlled mineralization is distinguished by a higher level of biological regulation. The organism actively controls minerals' nucleation, growth, and spatial organization in this mineralization. The organism can precisely regulate the formation and arrangement of mineral phases, producing well-structured and species-specific biominerals. This regulated activity often occurs inside specific intracellular compartments or matrices (Gadd 2010).

4.4 Bioleaching

Bioleaching is a biotechnological process that utilizes the metabolic activities of chemolithotrophic bacteria to convert insoluble metal sulfides, especially those containing arsenic, into soluble metal sulfates. The term used to describe this process is bioleaching. Microorganisms can transform solid substances into recoverable forms, and our approach exploits that potential. Several diverse pathways may be used for the arsenic bioleaching process. These processes include the direct transformation of solid-phase As(V) into the more mobile As(III), the oxidation of sulfides containing arsenic, or the dissolution of iron hydroxides caused by reduction. Moreover, conditions that promote dissimilatory reduction might enhance the movement of arsenic in the environment, thus triggering the liberation of arsenic from polluted soils. This is achieved by inducing the disintegration of the sorbing phases. This mechanism plays a significant role in the arsenic biogeochemical cycle and demonstrates the potential of microorganisms to help extract and retrieve metal compounds (Sher and Rehman 2019).

4.5 Biosurfactant Production

Biosurfactants are a group of amphiphilic biomolecules produced by many microbes, known for their ability to interact with surfaces. These compounds have significant variability in their structure, consisting of both hydrophilic (polar) and hydrophobic (nonpolar) moieties. Biosurfactants possess a distinctive amphipathic property that allows them to efficiently reduce surface tension and improve the solubility of hydrophobic substances. This makes them highly useful in various applications, including bioremediation, enhanced oil recovery, and the creation of environmentally friendly detergents and emulsifiers. Biosurfactants have attracted considerable interest due to their adaptability and ecologically friendly properties, making them attractive substitutes for synthetic surfactants in several industrial and medicinal fields (Mulligan 2021). Rhynolipid, a specific kind of glycolipid biosurfactant, is often used to eliminate a diverse range of heavy metals, namely arsenic. Biosurfactants often interact with heavy metals at the interface between soil and water, forming stable complexes. This interaction ultimately facilitates desorption by reducing the interfacial tension between the soil and the water. Consequently, the movement of heavy metals throughout the environment is enhanced (Mulligan 2021).

4.6 Biovolatilization

Biovoltaics uses microorganisms' catalytic activity of microorganisms to convert toxic heavy metals into their volatile forms. Arsenic methyltransferase enzymes, namely ArsM and AS3MT, facilitate the biotransformation of arsenic by converting arsenate (As(V)) into various methylated arsenic compounds, including monomethylated, dimethylated and trimethylated species. The last step in this enzymatic process synthesizes the volatile compound trimethylarsine (TMAIII). The biovolatilization of arsenic offers an effective strategy to control arsenic pollution by decreasing its accumulation in the food chain (Savage et al. 2018). Nevertheless, this process has intrinsic constraints since transporting volatile methylated arsenic species across large distances via the atmosphere and their subsequent precipitation might lead to the pollution of previously untouched regions or the re-deposition of arsenic at the initial locations. Furthermore, the intermediate by-products generated during the arsenic methylation, namely monomethylarsonic acid (MMAIII) and dimethylarsinic acid (DMAIII), are considered to be more harmful than other forms of arsenic. Hence, it is crucial to thoroughly grasp the environmental fate and possible dangers of biovolatilization as a remediation method.

4.7 Microbe-Assisted Phytoremediation

Phytoremediation is an environmentally friendly and renewable biotechnological approach that uses the inherent capabilities of plants to decrease the levels of pollutants or mitigate their harmful effects in various environmental settings. Phytoremediation involves many specialized procedures, such as phytoextraction, rhizofiltration (or phytofiltration), phytostabilization, and phytovolatilization (Sandhi et al. 2018).

4.7.1 Phytoextraction

Phytoextraction, often referred to as phytoaccumulation, phytoabsorption, or phytosequestration, is a bioremediation technique in which plants absorb pollutants from the soil or water through their roots and then transfer these compounds to their above-ground biomass, particularly in shoots. Phytoextraction is a term that is often used interchangeably with phytoaccumulation. Due to this physiological process, plants can effectively accumulate and isolate toxins inside their above-ground structures, thus facilitating the elimination of dangerous substances from contaminated environments (Rafati et al. 2011).

4.7.2 Phytofiltration

Phytofiltration is a novel remediation approach that involves using plants to remove pollutants, such as arsenic, from contaminated surface waterways or wastewater. Rhizofiltration, blastofiltration, and caulofiltration are examples of plant structures used in this strategy. Rhizofiltration utilizes roots, blastofiltration involves seedlings, and caulofiltration employs excised shoots. Phytofiltration utilizes plant components' natural absorption and adsorption characteristics to effectively cleanse aquatic environments and decrease the presence of hazardous contaminants (Sandhi et al. 2018).

4.7.3 Phytostabilization

Phytostabilization, or phytoimmobilization, is a remediation technique that involves carefully choosing and deploying certain plant species to stabilize pollutants within contaminated soils. This strategy aims to decrease the movement and availability of contaminants, thus reducing the probability of these pollutants migrating into groundwater resources and minimizing the risks of environmental contamination associated with this migration. Plants can efficiently immobilize heavy metals in soils using many methods, such as root sorption, precipitation, complexation, and the reduction of metal valence in the rhizosphere. Phytostabilization effectively prevents the buildup of heavy metals in living organisms and minimizes their release into groundwater by using these methods. It is crucial to understand that phytostabilization does not provide a permanent solution since heavy metals stay confined in the soil matrix. Instead, its movement is restricted rather than completely removed, which means there is still a possibility of future environmental hazards (Wuana and Okieimen 2011).

4.7.4 Phytovolatilization

Phytovolatilization is a method that enables genetically modified plants to absorb specific metal contaminants, such as arsenic (As), from the soil. Subsequently, these metal pollutants transform into gaseous states within the facility and are ultimately emitted into the atmosphere. Even with the recognition of this process as a potential resistance strategy in marine algae, the ability of terrestrial plants to significantly volatilize arsenic remains uncertain. Research examining the digestion and distribution of arsenic in taller plants found that the predominant accumulation occurs in the roots, with only a small amount transported to the shoots. However, plants can accelerate the process of arsenic biotransformation through their interactions with rhizospheric bacteria, potentially enhancing the volatilization rates (Savage et al. 2018).

The benefits and drawbacks of different arsenic remediation have been summarized in Table 4.3.

4.8 Future Prospects

Further investigation of the specific mechanisms and genetic factors that facilitate the phytoremediation of arsenic (As) is necessary despite increasing knowledge of the interactions between microorganisms and plants. In the future, it is essential for research to prioritize functional assays to determine the advantages that bacteria provide to plants under arsenic stress. Although polygenic markers such as 16S rRNA can provide valuable information on invading bacteria, excessive dependence on phylogenetic analysis may obscure the functional importance of mobile genetic components in the inherently complex and dynamic plant microbiome. Therefore, implementing a new breeding and genetics initiative that prioritizes the role of the microbiome in arsenic phytoremediation should provide significant benefits. Moreover, to progress in practical applications, particularly in bioenergy generation and microbial fuel cell technology, it is crucial to possess a comprehensive understanding of the processes included in the respiration of arsenic by bacteria (Irshad et al. 2021).

Notwithstanding the considerable potential of phytoextraction using hyperaccumulators, the fate of the gathered biomass remains a significant concern. Composting, pyrolysis, and biogasification are feasible solutions to this problem. Although built wetlands (CWs) provide a sophisticated method of remediating soil and water polluted with arsenic, it is crucial to use a comprehensive approach, especially in temperate or cold areas. Given that these materials have been extensively investigated for other contaminants but not specifically for arsenic, it is imperative to study unconventional wetland substrates. Notable examples of these substrates include waste generated from industrial and agricultural activities that possess substantial sorption capacities. A thorough cost–benefit analysis is crucial for the practical use of bioremediation protocols. Therefore, this will ensure that the methods are both financially viable and effective concerning their environmental impact (Dokania et al. 2024).

Table 4.3 Comparative analysis of the benefits and drawbacks of different As bioremediation systems (Rahman and Singh 2020; Sandhi et al. 2018; Savage et al. 2018; Yin et al. 2022)

Technology	Advantages	Disadvantages
Biosorption	The process reversibility Regenerability of biosorbents	The pH range may be a restriction as most heavy metals biosorb between 2.0 and 7.0 Sorbent binding sites are non-specific and may bind trace elements and THMs
Bioaccumulation	Highly efficient	Heavy metal buildup in biological systems may harm cellular processes and cause health problems As heavy metal concentrations rise, cells may engage efflux systems to discharge them into the environment
Biotransformation	Highly efficient Easy to use Ex-situ	As (III) exhibits a significant amount of toxicity Limited partial oxidation occurs at 50–200 μg/l The interaction between bacteria and their minerals reduces iron mineral adsorption
Biomineralization	Simple to operate in situ	Leaching mineralized phases might result in the potential for secondary contamination in the environment
Bioprecipitation		
Bioleaching	Economically feasible and environmentally friendly More efficient than chemical leaching	High concentrations of heavy metals significantly hinder the growth and metabolic activity of sensitive microorganisms In environments contaminated with pollutants, elevated levels of solid and organic matter may exacerbate these inhibitory effects, therefore diminishing the efficacy of bioleaching by microorganisms
Biosurfactant production	The capacity to operate well at elevated pH levels Excellent biodegradability with little toxicity	High production cost
Biovolatilization	Control As pollution by avoiding food chain buildup	During long-term transit and atmospheric precipitation, arsenic has the potential to disperse to uncontaminated regions or settle in their initial locations MMAIII and DMAIII, As methylation intermediates, are more hazardous than other As forms
Phytoextraction	Remediate huge areas of shallow contaminated soil with low to moderate metal contaminants	Possible metal leakage to groundwater A lack of thorough research on metal-chelating agent complexes in polluted soils

(continued)

Table 4.3 (continued)

Technology	Advantages	Disadvantages
Phytofiltration	Efficient management Economical Less need for secondary waste disposal	Few Hyperaccumulators—plants with high As accumulation—have been found
Phytostabilization	Particularly successful on fine-textured soils that have a high concentration of organic matter Applied to many sites with substantial surface pollution Cost-effective Uncomplicated Aesthetically pleasing	All heavy metals are retained in the soil; their mobility is restricted
Phytovolatilization	Control As pollution by avoiding food chain buildup	During long-term transit and atmospheric precipitation, arsenic has the potential to disperse to uncontaminated regions or settle in their initial locations MMAIII and DMAIII, As methylation intermediates, are more hazardous than other As forms

4.9 Conclusion

Microorganisms, particularly bacteria of several genera, have evolved into very effective bioremediation for eliminating heavy metals, such as arsenic (As), through various biochemical processes. Through their ability to degrade or transform arsenic, these microorganisms effectively alleviate the detrimental impacts of arsenic in contaminated environments. These organisms' metabolic pathways and enzymatic capacities are crucial in detoxifying toxic substances, thus significantly influencing the collective efficacy of bioremediation techniques. Optimizing the elimination of arsenic from polluted water sources requires a comprehensive understanding of the specific strategies used by these bacterial communities. Recent experimental findings have shown that the biological removal of arsenic may be effectively achieved under optimized pH, temperature, and oxygen availability. Many bacterial processes actively detoxify arsenic, including transformation, leaching, accumulation, and methylation. Despite the ongoing investigation of the effectiveness of these biological cleanup procedures, phytoremediation has gained considerable traction, especially with the advent of advanced technologies such as PS-S. Integration of arsenic-resistant plant growth-promoting microorganisms has potential.

In contrast, the complexity and time-consuming character of biologically mediated processes might lead to reduced predictability compared to conventional repair methods, affecting the reliability of these processes. Particularly in soil and water/sediment systems contaminated with toxic heavy metals, the microbial processes crucial for heavy metal removal heavily depend on specific environmental conditions. Therefore, more research is necessary to elucidate the particular mechanisms and genetic elements that enable the phytoremediation of arsenic. This reality persists despite the growing comprehensiveness of our understanding of the interactions between microorganisms and plants. Furthermore, to successfully implement bioremediation practices, it is crucial to have a comprehensive understanding of the sources, composition, and potential hazards related to hazardous heavy metals found in polluted soils. To ensure the successful execution of bioremediation projects, it will be imperative to carry out thorough cost–benefit studies. Adopting this complete strategy would significantly enhance effective and environmentally friendly development, aiming to mitigate arsenic contamination in various ecological settings.

Acknowledgements The authors would like to express our gratitude to the Indian Institute of Technology Delhi and the Indian Institute of Technology Guwahati for providing the infrastructural support needed to proceed with the work.

References

Bhat, Abhijnan, Kamna Ravi, Furong Tian, and Baljit Singh. 2024. "Arsenic Contamination Needs Serious Attention: An Opinion and Global Scenario." *Pollutants* 4(2):196–211. https://doi.org/10.3390/pollutants4020013

Dokania, Puja, Sourav Maity, Pritam Bajirao Patil, and Angana Sarkar. 2024. "Isothermal and Kinetics Modeling Approach for the Bioremediation of Potentially Toxic Trace Metal Ions Using a Novel Biosorbent Acalypha Wilkesiana (Copperleaf) Leaves." *Applied Biochemistry and Biotechnology* 196(5):2487–2517. https://doi.org/10.1007/s12010-023-04678-5

Gadd, Geoffrey Michael. 2010. "Metals, Minerals and Microbes: Geomicrobiology and Bioremediation." *Microbiology* 156(3):609–43. https://doi.org/10.1099/mic.0.037143-0

Hubadillah, Siti Khadijah, Mohd Hafiz Dzarfan Othman, Paran Gani, Norshuhaila Mohamed Sunar, Zhong Sheng Tai, Khong Nee Koo, Mohamad Arif Budiman Pauzan, Nurul Jannah Ismail, and S. M. Shahrul Nizan Shikh Zahari. 2020. "Integrated Green Membrane Distillation-Microalgae Bioremediation for Arsenic Removal from Pengorak River Kuantan, Malaysia." *Chemical Engineering and Processing-Process Intensification* 153:107996. https://doi.org/10.1016/j.cep.2020.107996

Irshad, Sana, Zuoming Xie, Sajid Mehmood, Asad Nawaz, Allah Ditta, and Qaisar Mahmood. 2021. "Insights into Conventional and Recent Technologies for Arsenic Bioremediation: A Systematic Review." *Environmental Science and Pollution Research* 28:18870–92. https://doi.org/10.1007/s11356-021-12487-8

Jimenez, Paul Andres Jimenez, Ximena Díaz, Marx Leandro Naves Silva, Anyela Vega, and Nilton Curi. 2023. "Assessing and Understanding Arsenic Contamination in Agricultural Soils and Lake Sediments from Papallacta Rural Parish, Northeastern Ecuador, via Ecotoxicology Factors, for Environmental Embasement." *Sustainability* 15(5):3951. https://doi.org/10.3390/su15053951

Limo, Marion J., Anna Sola-Rabada, Estefania Boix, Veeranjaneyulu Thota, Zayd C. Westcott, Valeria Puddu, and Carole C. Perry. 2018. "Interactions between Metal Oxides and Biomolecules: From Fundamental Understanding to Applications." *Chemical Reviews* 118(22):11118–93. https://doi.org/10.1021/acs.chemrev.7b00660

van Lis, Robert, Wolfgang Nitschke, Simon Duval, and Barbara Schoepp-Cothenet. 2013. "Arsenics as Bioenergetic Substrates." *Biochimica et Biophysica Acta (BBA)-Bioenergetics* 1827(2):176–88.

Maity, S., Biswas, R. and Sarkar, A. 2020. Comparative valuation of groundwater quality parameters in Bhojpur, Bihar for arsenic risk assessment. Chemosphere, 259, 127398. https://doi.org/10.1016/j.chemosphere.2020.127398

Maity, S., Biswas, R., Verma, S. K. and Sarkar, A. 2021. Natural polysaccharides as potential biosorbents for heavy metal removal. In Food, medical, and environmental applications of polysaccharides, 627–665. Elsevier. https://doi.org/10.1016/B978-0-12-819239-9.00012-9

Maity, S., Nanda, S. and Sarkar, A. 2021. Colocasia esculenta stem as novel biosorbent for potentially toxic metals removal from aqueous system. Environmental Science and Pollution Research, 28(42), 58885–58901. https://doi.org/10.1007/s11356-021-13026-1

Maity, S., Patil, P. B., SenSharma, S. and Sarkar, A. 2022. Bioremediation of heavy metals from the aqueous environment using Artocarpus heterophyllus (jackfruit) seed as a novel biosorbent. Chemosphere, 307, 136115. https://doi.org/10.1016/j.chemosphere.2022.136115

Maity, S., Sarkar, D., Poddar, K., Patil, P. and Sarkar, A. 2023. Biofilm-mediated heavy metal removal from aqueous system by multi-metal-resistant bacterial strain Bacillus sp. GH-s29. Applied Biochemistry and Biotechnology, 195(8), 4832–4850. https://doi.org/10.1007/s12010-022-04288-7

Maity, S., Bose, S., Dokania, P., Lohar, S. and Sarkar, A. 2024. A comprehensive review of arsenic contamination in India with an emphasis on its detection through biosensors and bioremediation from the aqueous system. Environmental Quality Management, 33(4), 427–457. https://doi.org/10.1002/tqem.22100

Maity, S., Dokania, P., Goenka, M., Rahul, S., Are, R. P. and Sarkar, A. 2024. Techno-economic feasibility and life cycle assessment analysis for a developed novel biosorbent-based arsenic bio-filter system. Environmental Geochemistry and Health, 46(3), 79. https://doi.org/10.1007/s10653-023-01839-7

Maity, S., Dokania, P., Goenka, M., Patil, P. B. and Sarkar, A. 2025. Health-Risk Assessment of Groundwater Arsenic Levels in Bhagalpur, India, and Development of a Cost-Effective Paper-Based Arsenic Testing-Kit. CLEAN–Soil, Air, Water, 53(1), p.e202300291. https://doi.org/10.1002/clen.202300291

Mujawar, Sajiya Yusuf, Kashif Shamim, Diviya Chandrakant Vaigankar, and Santosh Kumar Dubey. 2019. "Arsenite Biotransformation and Bioaccumulation by Klebsiella Pneumoniae Strain SSSW7 Possessing Arsenite Oxidase (AioA) Gene." *Biometals* 32:65–76. https://doi.org/10.1007/s10534-018-0158-7

Mulligan, Catherine N. 2021. "Sustainable Remediation of Contaminated Soil Using Biosurfactants." *Frontiers in Bioengineering and Biotechnology* 9:635196. https://doi.org/10.3389/fbioe.2021.635196

Nurchi, Valeria M., Aleksandra Buha Djordjevic, Guido Crisponi, Jan Alexander, Geir Bjørklund, and Jan Aaseth. 2020. "Arsenic Toxicity: Molecular Targets and Therapeutic Agents." *Biomolecules* 10(2):235. https://doi.org/10.3390/biom10020235

Pandey, Dheeraj, Harbans Kaur Kehri, Ifra Zoomi, Ovaid Akhtar, and Shweta Chaturvedi. 2021. "Bioremediation of Arsenic Contamination from the Environment: New Approach to Sustainable Resource Management." *Journal of Applied and Natural Science* 13(4):1499–1517. https://doi.org/10.31018/jans.v13i4.2986

Preetha, Jaganathan Sakthi Yazhini, Muthukrishnan Arun, Nandakumar Vidya, Kumaresan Kowsalya, Jayachandran Halka, and Gabrijel Ondrasek. 2023. "Biotechnology Advances in Bioremediation of Arsenic: A Review." *Molecules* 28(3):1474. https://doi.org/10.3390/molecules28031474

Rafati, M., N. Khorasani, F. Moattar, A. Shirvany, F. Moraghebi, and S. Hosseinzadeh. 2011. "Phytoremediation Potential of Populus Alba and Morus Alba for Cadmium, Chromuim and Nickel Absorption from Polluted Soil." *International Journal of Environmental Research* 5(4):961–70. https://doi.org/10.22059/ijer.2011.453

Rahman, Zeeshanur and Ved Pal Singh. 2020. "Bioremediation of Toxic Heavy Metals (THMs) Contaminated Sites: Concepts, Applications and Challenges." *Environmental Science and Pollution Research* 27(22):27563–81. https://doi.org/10.1007/s11356-020-08903-0

Roy, Vivek, Barnan Kumar Saha, Samarpita Adhikary, Madhumita G. Chaki, Monalisha Sarkar, and Ayon Pal. 2024. "Isolation, Characterization, Identification, Genomics and Analyses of Bioaccumulation and Biosorption Potential of Two Arsenic-Resistant Bacteria Obtained from Natural Environments." *Scientific Reports* 14(1):5716. https://doi.org/10.1038/s41598-024-56082-6

Roy, M., C. M. van Genuchten, L. C. Rietveld, and D. van Halem. 2023. "Integrating Biological Oxidation of Arsenite by Arsenic Oxidizing Bacteria with Iron-Electrocoagulation: A Novel Approach for Enhanced Removal of Arsenite from Water." Pp. 388–89 in *Arsenic in the Environment: Bridging Science to Practice for Sustainable Development As2021*. CRC Press. https://doi.org/10.1201/9781003317395-158

Sandhi, Arifin, Tommy Landberg, and Maria Greger. 2018. "Phytofiltration of Arsenic by Aquatic Moss (Warnstorfia Fluitans)." *Environmental Pollution* 237:1098–1105.

Sanyal, Tamalika, Pritha Bhattacharjee, Somnath Paul, and Pritha Bhattacharjee. 2020. "Recent Advances in Arsenic Research: Significance of Differential Susceptibility and Sustainable Strategies for Mitigation." *Frontiers in Public Health* 8:464. https://doi.org/10.3389/fpubh.2020.00464

Savage, Laurie, Manus Carey, Paul N. Williams, and Andrew A. Meharg. 2018. "Biovolatilization of Arsenic as Arsines from Seawater." *Environmental Science & Technology* 52(7):3968–74. https://doi.org/10.1021/acs.est.7b06456

Sher, Shahid and Abdul Rehman. 2019. "Use of Heavy Metals Resistant Bacteria—a Strategy for Arsenic Bioremediation." *Applied Microbiology and Biotechnology* 103:6007–21. https://doi.org/10.1007/s00253-019-09933-6

Simon, M., Joshi, H., Yadav, A. K., & Hopke, P. K. (2024). Simultaneous nitrogen and phosphorus removal from polluted water using a novel bacterial consortium (BACON-38): Promising application for wastewater remediation. Journal of Water Process Engineering, 66, 106044. https://doi.org/10.1016/j.jwpe.2024.106044

William, Vladimir U. and Hilbert D. Magpantay. 2023. "Arsenic and Microorganisms: Genes, Molecular Mechanisms, and Recent Advances in Microbial Arsenic Bioremediation." *Microorganisms* 12(1):74. https://doi.org/10.3390/microorganisms12010074

Wuana, Raymond A. and Felix E. Okieimen. 2011. "Heavy Metals in Contaminated Soils: A Review of Sources, Chemistry, Risks and Best Available Strategies for Remediation." *International Scholarly Research Notices* 2011(1):402647. https://doi.org/10.5402/2011/402647

Yadav, A. K., Sahoo, S. K., Mahapatra, S., Kumar, A. V., Pandey, G., Lenka, P., & Tripathi, R. M. (2014). Concentrations of uranium in drinking water and cumulative, age-dependent radiation doses in four districts of Uttar Pradesh, India. Toxicological & Environmental Chemistry, 96(2), 192-200. https://doi.org/10.1080/02772248.2014.934247

Yin, Shiqian, Xuan Zhang, Huaqun Yin, and Xian Zhang. 2022. "Current Knowledge on Molecular Mechanisms of Microorganism-Mediated Bioremediation for Arsenic Contamination: A Review." *Microbiological Research* 258:126990. https://doi.org/10.1016/j.micres.2022.126990

Chapter 5
Integrated Bioremediation Technologies for Arsenic Removal

Abhay B. Fulke and Siddant Ratanpal

Abstract Contamination caused by Arsenic (As) in soil and water poses a significant threat to overall health of environment and humans. Bioremediation of As makes use of bacteria, fungi, and yeast for detoxification/removal of As from the contaminated site. Phytoremediation, another conventional method, employs plants to absorb, sequester, and detoxify arsenic from contaminated environments. This chapter explores various bioremediation approaches to mitigate arsenic pollution in the environment. The chapter delves into the molecular mechanisms underlying arsenic bioremediation. Key processes such as redox reduction mechanisms, the action of arsenate reductase and arsenite oxidase enzymes, and methylation are discussed in detail. Understanding the mechanism of converting arsenate [As (V)] to arsenite [As (III)] and its further removal from biological systems is essential. Recent advances in bioremediation technologies are also mentioned in this chapter, which are essential so as increase the efficiency, with a focus on integrated techniques that enhance arsenic removal efficiency. Genetically modified organisms (GMOs) have shown a promising improvement in arsenic uptake and detoxification capabilities of microbial and plant systems. Integrated bioremediation approaches, such as phytobial bioremediation, combine the strengths of both microbial and phytoremediation techniques for a synergistic effect.

Keywords Arsenic contamination · Bioremediation · Phytoremediation · Arsenic removal · Phytosuction · Arsenotrophy

A. B. Fulke (✉)
Microbiology Division, Regional Centre, CSIR-National Institute of Oceanography (CSIR-NIO), Andheri (West), Mumbai, Maharashtra, India
e-mail: afulke@nio.res.in

Academy of Scientific and Innovative Research (AcSIR), Ghaziabad, India

S. Ratanpal
Amity Institute of Biotechnology, Amity University Maharashtra, Mumbai, India

A. K. Yadav et al. (eds.), *Arsenic Removal Technologies*,
https://doi.org/10.1007/978-3-031-91171-2_5

5.1 Introduction

Arsenic (As) is placed 20th in abundance parameter inside the earth's crust, with a concentration of 1.7 to 1.8 mg/kg. As is a metalloid that is widely dispersed in rock, soil, and water systems. As (III) (arsenite) and As (V) (arsenate) are the most abundant forms of As found in nature. Both kinds of As are essential inorganic elements. Under oxidative circumstances, As (V) predominates, while As (III) does so under reductive ones. Although As may exist in a range of oxidation states, natural waters typically feature As(III) or As(V), depending on the environment's redox state. As may react with carbonate ligands to produce liquid complexes, and As (III) can react with reduced sulfur to make thioarsenites, which may be present in certain groundwater. The water present on the surface and even soil solutions high in organic matter may include organic As species produced by microbial methylation activities, although groundwater seldom contains a considerable number of these species. Metalloid As is often used in agricultural pesticides, preservatives for the wood and glass industries, herbicides, and other applications. The combined influence of human and natural variables can result in large fluctuations in As concentrations. As can be discovered or released naturally through processes such as rock weathering and volcanic eruptions. As a result of these processes, As can poison the environment. As noted previously, it enters the ecosystem through the soil and water systems and can accumulate in plants, fish, and humans (Gupta et al. 2024). When As enters the human body, it can cause a myriad of acute, chronic, and lethal health problems and diseases, which may or may not have specific symptoms, depending on the concentration, chemical form, and exposure period. As exposure and its consecutive poisoning can result in a myriad of symptoms including fatigue, weight loss, general weakness, anorexia, hair loss, and other indicators, while As-associated illnesses and disorders are linked to liver and kidney disorders, cardiovascular, peripheral vascular function, and hyperpigmentation (Bhuvaneswari et al. 2024). The concentration of arsenic in soil varies depending on the available properties of the parent materials available; arsenic is often found in the baseline soil at levels ranging from 5 to 10 mg/kg. Arsenic values in Asian nations typically range from 5 to more than 50 mg per kilogram. Arsenic concentrations in bog and peat soils are quite high, reaching 13 mg/kg. Sulphate soils have the highest amounts of arsenic (Al-Makishah 2020). Arsenic concentrations vary widely due to the diverse parent minerals found in different soils, however, the United States Environmental Protection Agency (USEPA) suggested tolerable arsenic value of 24 mg/kg has been surpassed due to various human exploration and industrial operations As eventually finds its way into plants via the soil. Arsenic (As) enters rice fields (Oryza sativa) through polluted agricultural water and past contamination of As-based fertilizers. Arsenite (As(III), arsenate (As(V) dimethylarsinic acid (DMA), and monomethylarsonic acid (MMA) were observed in plant absorption from a greenhouse study in which rice was watered with arsenate As(V) polluted water. As was first detected over 100 years ago in marine samples and it was quickly determined that commonly consumed seafoods such as fish, crabs, and mollusks contained exceptionally high amounts of As. Seafoods including these were found to have As species such as iAs and oAs. iAs have been observed in two oxidation states: arsenite (As-III) and arsenate (As-V); nevertheless, these forms are found in smaller amounts in seafood than other As molecules (Kato et al. 2020).

Thus, this chapter aims to provide an understanding of As bioremediation methods using bacteria, fungi, and algae, providing a complete understanding of the mechanisms involved, especially at molecular levels, and the enzymes that play a key role in conversion of toxic As. In addition, this chapter also explains techniques and advancements in As remediation such as phytosuction, genetically modified organisms for remediation, construction of wetlands, etc. These techniques will create a foundation for a more sustainable and efficient bioremediation of As.

5.2 Conventional Bioremediation Strategies for As Removal

5.2.1 Microbial

5.2.1.1 Bacterial

Bacteria play an important role in mobilization, transformation, and bioremediation through complicated metabolic processes such as reduction, oxidation (Bhakat et al. 2019), biosorption, methylation, and volatilization. Bacterial bioremediation of As occurs via processes such as bioaccumulation, surface level absorption (also known as biosorption), conversion of As (III) into trimethylarsine gas. Bacteria, such as *Aliihoeflea sp. and Achromobacter sp.*, have been shown to oxidize As (III) to less harmful As(V), and this technique appears to be a potential alternative for groundwater remediation (Biswas et al. 2019). Bacterial oxidation of As(III) is related to a catalytic function of an enzyme, As(III) oxidase, whose gene is labelled *aox* (Fig. 5.1). However, certain strains decrease As(V) to As(III). As a result, the use of bacterial bioremediation as a general remediation option is hampered by this hostile activity. Bacteria are investigated for bioremediation in the laboratory under sterile and optimal settings, but the synergistic action remains a problematic hypothesis in the natural environment, where many other microbial species exist.

5.2.1.2 Fungi and Yeast

According to research, isolated fungal species primarily aid in the bioremediation of polluted environments through biovolatilization, biosorption, and biotransformation. However, some research on yeast has shown that it has the potential to be employed in As bioremediation. In yeast, the primary genes involved in the resistance mechanism of resistance and elimination are *acr1,2 and 3.* These particular genes are known to code for the arsenate reductase enzyme, which is well known for converting arsenate

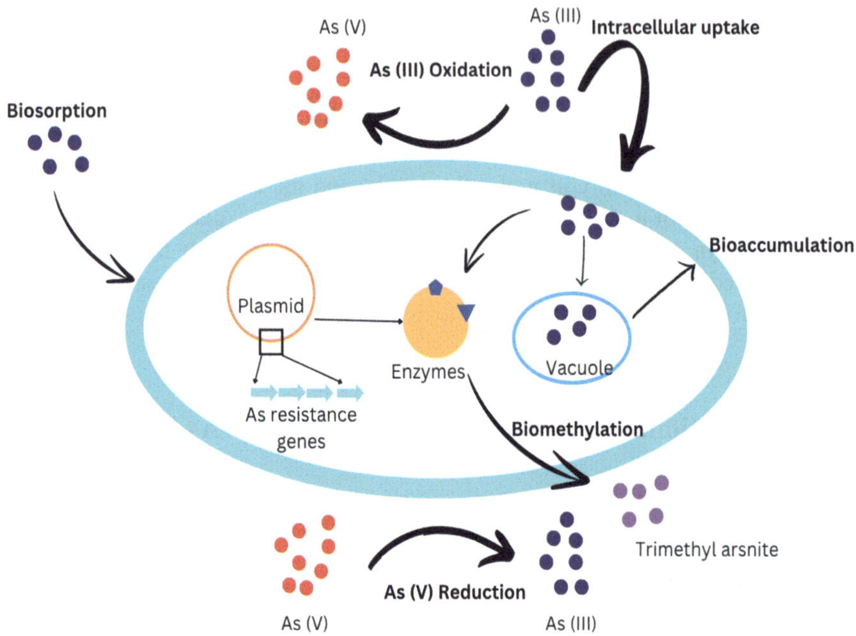

Fig. 5.1 As removal and reduction mechanisms by bacteria

to arsenite and subsequently eliminating it from the cell machinery using *acr3,* which is a transporter gene (Mukhopadhyay 2000).

5.2.1.3 Algal

The ability of algae, either microalgae or macroalgae, to absorb heavy metals has been widely accepted, especially in natural settings such as lakes and oceans and other water bodies (Tombokan et al. 2020). Several investigations have indicated that several algae species, for example, *Pithophora sp., Phormidium sp.*, and *Chlorella pyrenoidosa*, have increased As bioaccumulation capacity, up to 760 mg kg^{-1} (Arora et al. 2017). These functional groups provide the negative charges to the cell wall, which attracts and absorbs additional metal ions. Algal species differ when different components of the cell wall are present, which are accountable for different functional groups. Proteins and polysaccharides have highly efficient cell wall components that participate in the creation of a wide range of functional groups. The walls of green algae cell include polysaccharides that increase heavy metal bioaccumulation through carboxyl and sulfate functional groups (Arora et al. 2017).

5.2.2 Phytoremediation

After microbes, plants are used for As remediation as well. As seen in the case of microbes, biosorption and accumulation are prominent methods for remediation. There are several ways in which phytoremediation is achieved, one of which is the property of plants to act as hyperaccumulators. In hyperaccumulation plants are able to withhold more than 1000 mg/kg of pollutant or as in this case. To consider a plant as a hyperaccumulator, both its bioaccumulation factor and its translocation factor are taken into account. Both of these factors should be greater than 1. *The Pteris* genus is the one in which most As hyperaccumulation has been observed and also consists of most of literature on it. It makes up 2% of biomass in *Pteris* genus and especially in *Pteris vittata.* In a pilot-scale investigation, *P. vittata* was employed to reduce As concentrations in drinking water using a continual phytofiltration system (Srivastava et al. 2021). Over three months, up to 1900 L/day of water that had an initial As content of 10.2 μg/L was remediated, revealing As values as low as 2 μg/L.

Plant roots take up pollutants and nutrients, along with water, and retain them in different parts of the plant. Plants that are able to accumulate metals up to 1% of their dry mass are called hyperaccumulators. The researchers investigated the As-accumulating capacity of native plants and discovered several As-collecting/hyperaccumulating native species.

Although As-accumulating plants acquire less As, their large biomass allows them to reach their target in less time (Srivastava et al. 2021), It also decreases the threat of invasion by alien fern species and lowers costs. Brassicas are excellent candidates for hyperaccumulating As and other heavy metals (Table 5.1).

Phytostabilization or phytorestoration is the utilization of a plant's broad root system and related rhizobiome to retain pollution at the site, minimizing leaching into groundwater and windblown exposure to air resulting from soil erosion. It is also useful for infiltration management, as it prevents significant volumes of precipitation from seeping into groundwater. Phytostabilization does not seek to eliminate As, but rather stabilize it at the site and reduce the risk of exposure (Bakshe and Jugade 2023).

Vegetation covering the soil aids in hydraulic management by impeding the horizontal and upward movement of pollutants into ground water. Another method of metal phytoremediation is to use naturally obtained or genetically modified plants

Table 5.1 List of plant hyperaccumulators

Plant species	References
Isatis cappadocica	Souri et al. (2017)
Chrysopogon zizaniodes	Gautam et al. (2017)
Cymbopogon flexuosus	Jha and Kumar (2017)
Azolla microphylla	Jha et al. (2015)
Eichhornia crassipes	Jha et al. (2015)

that can function alone or in synergy with rhizospheric bacteria, and remove As from the soil, physiologically converting it to volatile forms within the plants itself, and discharge it into the atmosphere (a process known as rhizofiltration). Although volatilization from natural habitats has been documented, the percentage of rhizomicrobes that lead to the general fraction remains unknown.

5.3 Molecular Mechanisms

5.3.1 Arsenite Oxidase

Arsenite oxidation takes place in aerobic and anaerobic environments, and the conversion of arsenite to arsenate is well recognized as an efficient technique for bioremediation of As. Arsenate is less toxic than As(III) and is also expected to be easily eliminated by precipitating using Al and Fe (Burton et al. 2014). Arsenite oxidase, especially AoxAB, which was earlier known as AioAB, is expected to catalyse the oxidation process of As(III) under aerobic circumstances.

It is composed of four domains: domain I binds the [3Fe-4S] cluster to the Rieske subunit together, domains. In aerobic conditions, the arsenite oxidation substrate, As(III), enters the enzyme surface and interacts with Mo(VI), which in turn is then converted to Mo(IV) by two-electron exchange, while the oxidative product As(V) leaves the pit at exactly the same time.

Unlike the Aox complex, Arx is species-specific and occurs in a subset of bacteria, most of which are closely related to the phylum *Proteobacteria*, specifically the Gammaproteobacteria class.

5.3.2 Arsenate Reductase

Arsenate reduction, as well as arsenite oxidation, is vital in As-polluted bioremediation. Because of its high solubility, As(III) preferentially moves from the soil to the aqueous phase, decreasing As concentrations in the soil along with crops. Usually, elimination of arsenate is overseen by As (V) reductases family, which are further divided into respirator and cytoplasmic Arsenate reductases. The *Staphylococcus aureus* plasmid pI258 has been reported to carry Arsenate reductase, which is responsible for the mitigation of Arsenate. Usually, the *ars* operon that encodes the Arsenate reductase complexes consists of three (ArsRBC) or five genes. In Escherichia coli, five genes were reported to be involved in plasmic-oriented As detoxification. These genes were Ars A, B, C, D, and R, while Staphylococcus only had three.

5.3.3 *Methylation*

Methylation of the inorganic form of As is regarded a crucial way of As bioremediation. In this process, As is converted from an inorganic form to a much less hazardous organ arsenic molecule, together with the production of volatile TMA gases (Satyapal et al. 2018). Every year, about 2.1×10^7 kg of As is released into the atmosphere from land and/or aquatic bodies, highlighting the need for As cleanup. Although trimethylated arsenic (TMA) is more dangerous than As(III), which is a methylated intermediary molecules, such volatile forms do not remain in cells. Furthermore, methylation compounds such as trimethyl arsine oxide (TMAO) are significantly less hazardous than As(III) (Verma et al. 2016). As methylation has been associated with a variety of microbial species throughout the years, including Pseudomonas spp., Aspergillus spp., and Penicillium spp. In general, As(V) must be reduced to As(III) before it can be methylated.

5.4 Recent Developments in Arsenic Modification

5.4.1 *Genetically Modified Organisms (GMO)*

Bioremediation is an environmentally friendly and inexpensive method that can be used to clean complicated industrial tannery effluent, including HMs, which poses a significant hazard to ecosystem pollution (Verma et al. 2021). Metal toxicity is a major worry due to its bioaccumulation in nature and its inability to biodegrade. GMOs have been used for the bioremediation of HM because they are environmentally friendly, cost-effective, simple, and scalable. Several parameters can impact the efficacy of GMO-based bioremediation, including reaction conditions, heavy metal chemical makeup, redox potential, and nutritional state.

In conditions relating to water and soil, the existence of arsM-expressing bacteria has been reported to significantly increase arsenic methylation. Thus, a volatilization strategy based on genetically modified bacteria (GM) has been presented to possess a potential As removal tool. The capacity of microorganisms to biovolatilize varies with a range of factors, such as genetic variation, soil temperature, and organic matter, influence As volatilization. As a result, genetically modified plants with increased methylation capacity are proposed for As bioremediation. For example, in a study conducted by *Staphylococcus sp*, which codes the *ars* gene, arsenic may absorb and at the same time volatilize it very efficiently. The E. coli strain containing the arsM gene which was cloned from bacteria *R. palustris* turned arsenic resistant and accelerated the creation of numerous intermediates from arsenite, such as TMA, finally ejecting the arsenic from the cellular system. Another example is GM *Bacillus subtilis* 168, which encodes the.

Cyanidioschyzon merola CmarsM gene and can remove As from organic manure. In a 48-h investigation in the same study, transformed *B. subtilis* 168 transformed the majority of the As present in the methylation medium through the process to methylated biological arsenic and was capable of volatilizing large amounts of DMA and TMA (Huang et al. 2015). When inoculated with As-contaminated organic manure, the mutant strain significantly increased As volatilization at 50 °C. Some key aspects that impact GEO bioremediation include sorption locations, the structure of the microbial cell wall, and ionization of chemical entities on the cell wall (Huang et al. 2015).

5.4.2 Integrated Approaches for Bioremediation (Phytobial Bioremediation)

Phytobial remediation integrates bioremediation and phytoremediation to minimize the target pollutant. The majority of bacteria on the planet live in soil. Soil bacteria play a critical role in ecosystems, driving significant biogeochemical processes, regulating plant production, and recycling nutrients. Soil microorganisms boost plant production in a habitat with limited nutrients by providing growth-limiting nutrients. They can also reduce plant viability by serving diseases. They are also known to convert nutrients into non-accessible forms, which in turn reduces plant productivity, or into accessible forms, improving productivity for plants (Fig. 5.2). Furthermore, bacteria, like other biosphere components, play an important role in As global cycle by engaging in rock weathering, mineral dissolution, and biotransformation events. Given the active function of the soil microbiota in the ecosystem, a phytoremediation program will not be successful without knowledge of the native microbial flora that predominates in the region.

As a result, including minerals does not qualify as a free test site. The site should strike an ideal equilibrium between its dissimilatory arsenate-remediating microbial biodiversity and a mixed phyto- and microbial flora that fishes or scavenges bioavailable As soon as it becomes available. The process of releasing and eating As would continue until all of the As-rich mineral resources were depleted to feed the bacterium. A site is termed free after all of the cached As has entered the solution, and no additional As species have invaded from invasive sources. In a system of nature, both bacteria and plants play a crucial role in healing the environment from the harmful impacts of pollutants such as As. Rahman et al. (2014) found that prokaryotes and eukaryotes play a crucial role in the As elimination process. In this work, we will investigate the unique roles and methods of free-living, rhizomic, and endomic microorganisms, as well as their interactions with plants throughout the As elimination process. Rhizosphere-associated bacteria are known as rhizobiomes, while root-associated mycorrizha form a separate subgroup known as mycorhizobiomes.

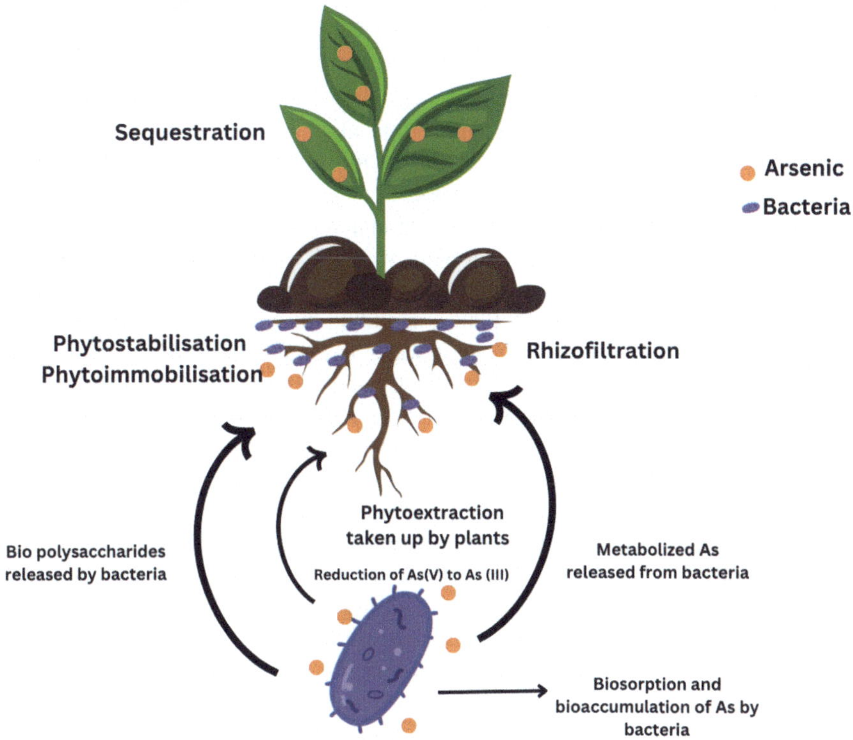

Fig. 5.2 Phytobial bioremediation of As from soil

5.4.3 Phytosuction Separation

Several reports have shown that phytoremediation is an inexpensive and ecologically sustainable method of mitigating toxic metal-polluted soils. However, the use of phytoextraction (PE) approach, even with hyperaccumulators, requires longer periods of time and is thus impracticable. Phytosuction separation is a revolutionary phytoremediation technique that was recently developed (Katoh et al. 2016). In this process, a chemical, such as ferrihydrite, is added to the contaminated soil to immobilize the desired harmful metal and/or metalloid, and plants grow on the surface of the previously applied material. Metals and metalloids are carried to and fixed in materials due to the phenomenon of water suction action of on growing plant roots. The flexibility of the target metal or metalloid in the study is the primary factor that determines the effectiveness of the PS-S method.

5.4.4 Constructed Wetlands

Previous studies have demonstrated that phytoremediation can eliminate As from polluted water bodies. The practical applications include the purification of polluted waters in constructed wetlands (CW). Constructed wetlands are places that mimic the environments found in natural wet locations (such as swamps) to remediate filthy water (Ekperusi et al. 2024). Contaminated water flow is present on the surface or underneath, and surface-flowing wetlands frequently need larger area than subterranean-flowing constructed wetlands; however, they can be simplified and inexpensive (Ekperusi et al. 2024). Plants, including those found in the surrounding environment, play a crucial role in wastewater treatment. Sandoval et al. (2019) discovered that plants in man-made wetlands offer oxygen, allowing toxins to be removed more quickly. Other advantages include lower affluent flow speed, improved sedimentation, and absorption of nutrients. Furthermore, the harvest or elimination of As-accumulating plants from these environments contributes to a reduction in the levels of water pollution. The plants utilized in CW have been better characterized. According to Arslan et al. (2019), the first genetic research on *Juncus effuses* was performed with the purpose of furthering research on natural and man-made wetlands. According to the author, genome sequencing refers to the analysis of the complete genetic makeup of plants used in constructed wetlands is complete, and extensive literature has been established. Other studies imply that constructed wetlands are effective arsenic removal options. A study by Thathong et al. (2019) examined the elimination of arsenic in a pilot phase CW with *Colocasia esculenta*. A full-scale examination of CW-based systems for unclean water treatment was also carried out.

5.4.5 Arsenotrophy

Despite its highly toxic nature, As is required by various microbes for its use in energy metabolization. The utilization usually involves either reduction or oxidization through electron transfer for nutritional needs or respirations (Anand et al. 2022). Three kinds of enzymes mainly *ArrA, AioA, and ArxA* are known to catalyze energy-conserving processes in biological systems in arsenic metabolism: These constitute different clades in the DMSO that is the reductase superfamily of enzymes containing molybdenum. In the periplasmic regions of bacteria, the enzymes ArrA, AioA, and ArxA tend to be soluble but interact with the membrane and subsequent membrane anchoring subunits, allowing the transfer of electrons for arsenate reduction or arsenite oxidation.

5.5 Conclusions

Arsenic (As) bioremediation represents a vital and evolving field that aims to address the significant threat posed by arsenic contamination to human and environmental and human health. This chapter has explored various bioremediation strategies, highlighting the pivotal roles of bacteria, fungi, yeast, and plants in detoxifying and removing arsenic from contaminated sites. Key molecular mechanisms such as redox reactions, enzymatic actions involving arsenate reductase and arsenite oxidase, and methylation processes have been discussed, underscoring their importance in understanding and enhancing bioremediation efficacy.

Recent advances in bioremediation technologies, including the use of genetically modified organisms (GMOs), have shown promising improvements in arsenic uptake and detoxification. Integrated bioremediation approaches, such as phytobial bioremediation, take advantage of the synergistic strengths of microbial and phytoremediation techniques, offering improved arsenic removal efficiencies. Innovative solutions such as phytosuction separation, constructed wetlands, and sequestration strategies further contribute to the arsenal of arsenic remediation methods. In addition, the concept of arsenotrophy, in which certain organisms use arsenic for growth and metabolism, introduces a novel and sustainable avenue for bioremediation. By harnessing these natural processes and advancing integrated bioremediation technologies, we can achieve more effective and sustainable solutions for the removal of arsenic. This comprehensive overview of bioremediation approaches for arsenic mitigation highlights the potential of these methods to significantly contribute to the protection of human and environmental health. Continued research and development in this field is essential to refine and optimize these technologies, ensuring that they can meet the growing challenges of arsenic contamination in a variety of environmental contexts.

Acknowledgements The authors express their heartfelt gratitude to our esteemed institute and all individuals who have played a role, directly or indirectly, in contributing to this study.

References

Al-Makishah, N.H., Taleb, M.A. and Barakat, M.A., 2020. Arsenic bioaccumulation in arsenic-contaminated soil: a review. *Chemical Papers*, *74*, pp. 2743–2757.

Anand, V., Kaur, J., Srivastava, S., Bist, V., Singh, P. and Srivastava, S., 2022. Arsenotrophy: A pragmatic approach for arsenic bioremediation. *Journal of Environmental Chemical Engineering*, *10*(3), p. 107528.

Arora N, Gulati K, Patel A, Pruthi PA, Poluri KM, Pruthi V (2017) A hybrid approach integrating arsenic detoxification with biodiesel production using oleaginous microalgae. Algal Res 24:29–39.

Arslan, M., Devisetty, U.K., Porsch, M., Große, I., Müller, J.A. and Michalski, S.G., 2019. RNA-Seq analysis of soft rush (Juncus effusus): transcriptome sequencing, de novo assembly, annotation, and polymorphism identification. *BMC genomics*, *20*, pp. 1–12.

Bakshe, P. and Jugade, R., 2023. Phytostabilization and rhizofiltration of toxic heavy metals by heavy metal accumulator plants for sustainable management of contaminated industrial sites: A comprehensive review. *Journal of Hazardous Materials Advances*, *10*, p. 100293.

Bhakat K, Chakraborty A, Islam E (2019) Characterization of arsenic oxidation and uranium bioremediation potential of arsenic resistant bacteria isolated from uranium ore. Environ Sci Pollut Res 26: 12907–12919.

Bhuvaneswari, Rajagopal, Narasimman Manickam, Ambrose Paulraj, and Razack Rafi Mohamed. "Heavy Metal Accumulation and Health Risk Assessment in the Puddles of Virudhunagar District, Tamil Nadu, India." In *Integrated Management of Water Resources in India: A Computational Approach: Optimizing for Sustainability and Planning*, pp. 173–194. Cham: Springer Nature Switzerland, 2024.

Biswas R, Vivekanand V, Saha A, Ghosh A, Sarkar A (2019) Arsenite oxidation by a facultative chemolithotrophic Delftia spp. BAs29 for its potential application in groundwater arsenic bioremediation. Int Biodeterior Biodegrad 136:55–62.

Burton, E.D., Johnston, S.G., Kocar, B.D., 2014. Arsenic mobility during flooding of contaminated soil: the effect of microbial sulfate reduction. Environ. Sci. Technol. 48, 13660–13667.

Ekperusi, A.O., Sikoki, F.D. and Nwachukwu, E.O., 2024. Sorption of cadmium, chromium, lead, and vanadium from artificial wetlands using Lemna aequinoctialis. *International Journal of Phytoremediation*, *26*(6), pp. 873–881.

Gao Z, Jiang C, Lyu R, Yang Z, Zhang T (2020) Optimization of the preparation of fungal-algal pellets for use in the remediation of arsenic-contaminated water. Environ Sci Pollut Res 27:3678936798.

Gautam M, Pandey D, Agrawal M (2017) Phytoremediation of metals using lemongrass (Cymbopogon citratus (DC) Stapf.) grown under different levels of red mud in soil amended with biowastes. Int J Phytoremediation 19:555–562.

Gunwal I, Singh L, Mago P (2014) Comparison of phytoremediation of cadmium and nickel from contaminated soil by Vetiveria zizanioides L. Int J Sci Res Publ 4:1–7.

Gupta, Sonam, Nivedita Mishra, Ankit Kumar, and Akhilesh Kumar Yadav. "Fluoride Pollution in Subsurface Water: Challenges and Opportunities." *Advanced Treatment Technologies for Fluoride Removal in Water: Water Purification* (2024): 19–39.

Hernandez-Maldonado, J., Sanchez-Sedillo, B., Stoneburner, B., Boren, A., Miller, L., McCann, S., Rosen, M., Oremland, R.S., Saltikov, C.W., 2017. The genetic basis of anoxygenic photosynthetic arsenite oxidation. Environ. Microbiol. 19, 130–141.

Huang K, Chen C, Shen Q, Rosen BP, Zhao FJ (2015) Genetically engineering Bacillus subtilis with a heat-resistant arsenite methyltransferase for bioremediation of arsenic-contaminated organic waste. Appl Environ Microbiol 81:6718–6724.

Jha AK, Kumar U (2017) Studies on removal of heavy metals by cymbopogon flexuosus. Int J Environ Agric Biotechnol 10:89–92.

Kato, L.S., Ferrari, R.G., Leite, J.V.M. and Conte-Junior, C.A., 2020. Arsenic in shellfish: A systematic review of its dynamics and potential health risks. *Marine Pollution Bulletin*, *161*, p. 111693.

Katoh, M., Hashimoto, K. and Sato, T., 2016. Lead and antimony removal from contaminated soil by phytoremediation combined with an immobilization material. *Clean–soil, air, water*, *44*(12), pp. 1717–1724.

Mateos, L.M., Villadangos, A.F., de la Rubia, A.G., Mourenza, A., Marcos-Pascual, L., Letek, M., Pedre, B., Messens, J., Gil, J.A., 2017. The arsenic detoxification system in corynebacteria: basis and application for bioremediation and redox control. Adv. Appl. Microbiol. 99, 103–137.

Rahman, S., Kim, K.H., Saha, S.K., Swaraz, A.M. and Paul, D.K., 2014. Review of remediation techniques for arsenic (As) contamination: a novel approach utilizing bio-organisms. *Journal of environmental management*, *134*, pp. 175–185.

Sandoval, L., Zamora-Castro, S.A., Vidal-Álvarez, M. and Marín-Muñiz, J.L., 2019. Role of wetland plants and use of ornamental flowering plants in constructed wetlands for wastewater treatment: A review. *Applied Sciences*, *9*(4), p. 685.

Satyapal, G.K., Mishra, S.K., Srivastava, A., Ranjan, R.K., Prakash, K., Haque, R., Kumar, N., 2018. Possible bioremediation of arsenic toxicity by isolating indigenous bacteria from the middle gangetic plain of bihar, India. Biotechnol. Rep. 17, 117–125.

Souri Z, Karimi N, Sandalio LM (2017) Arsenic hyperaccumulation strategies: an overview. Front Cell Dev Biol 5:67.

Srivastava, S., Shukla, A., Rajput, V.D., Kumar, K., Minkina, T., Mandzhieva, S., Shmaraeva, A. and Suprasanna, P., 2021. Arsenic remediation through sustainable phytoremediation approaches. *Minerals*, *11*(9), p. 936.

Thathong, V., Tantamsapya, N., Yossapol, C., Liao, C.H., Wirojanagud, W. and Padungthon, S., 2019. Role of Colocasia esculenta L. schott in arsenic removal by a pilot-scale constructed wetland filled with laterite soil. *Heliyon*, *5*(2).

Tombokan JL, Kepel RC, Mantiri DM, Paulus JJ, Lumingas LJ (2020) Comparison of seaweed communities in coastal waters with different heavy metals concentrations in Minahasa Peninsula, North Sulawesi, Indonesia. AACL Bioflux 13(4).

Verma, S., Verma, P.K., Meher, A.K., Dwivedi, S., Bansiwal, A.K., Pande, V., Srivastava, P.K., Verma, P.C., Tripathi, R.D., Chakrabarty, D., 2016. A novel arsenic methyltransferase gene of westerdykella aurantiaca isolated from arsenic contaminated soil: phylogenetic, physiological, and biochemical studies and its role in arsenic bioremediation. Metallomics 8, 344–353.

Verma, S., Bhatt, P., Verma, A., Mudila, H., Prasher, P. and Rene, E.R., 2021. Microbial technologies for heavy metal remediation: effect of process conditions and current spractices. *Clean Technologies and Environmental Policy*, pp. 1–23.

Chapter 6
Nanotechnology-Based Solutions: Advancing Efficiency and Affordability

Dev Raj, Alok Kumar, Ashutosh Kumar, Karan Kumar Paswan, and Kamal Nayan

Abstract Today, environmental pollution is continuously increasing and is becoming a threat to human life due to rapid industrial expansion and population growth. Contamination of groundwater with Arsenic (As) is becoming a global concern because of its lethal effects on human health. To address this issue, several technologies effectively remove Arsenic from groundwater (i.e., Coagulation, Ion-Exchange, Membrane, and Adsorption). Recently, a nanotechnology-based viable method has introduced for removing arsenic from the soil and water because of its robust, energy-efficient, affordable, and eco-friendly nature. Furthermore, the current advancement in nanotechnology-based solutions is being driven by mostly two techniques: (i) Absorption-based and (ii) Membrane-based. Nanomaterials are often employed for the purpose of adsorbents, since they have a huge surface area/volume ratio, which enhances the adsorption capability and renders them potentially appropriate for use in applications requiring greater removal efficiency. Several nanomaterials have recently been explored which can be categorized more specifically as metal, metal oxide-based nanoadsorbent, and carbon-derived nanoadsorbent. The nanotechnology membrane-based solution is another affordable method for eliminating the arsenic content from water up to a certain level. This chapter reviewed the application of different nanoadsorbents and membranes, their synthesis, properties, efficiency, limitations, and mechanism for the removal of arsenic from contaminated groundwater. Finally, future trends and perspectives are also discussed.

Keywords Nanomaterials · Arsenic · Nanoadsorbent · Adsorption · Membrane

---

D. Raj (✉) · A. Kumar · K. K. Paswan
Department of Chemical Engineering and Biochemical Engineering, Rajiv Gandhi Institute of Petroleum Technology, Jais, Amethi, India
e-mail: eche19018@rgipt.ac.in

A. Kumar
Department of Petroleum Engineering and Geo Engineering, Rajiv Gandhi Institute of Petroleum Technology, Jais, Amethi, India

K. Nayan
Department of Chemical Engineering and Applied Chemistry, University of Toronto, Toronto, ON, Canada

A. K. Yadav et al. (eds.), *Arsenic Removal Technologies*,
https://doi.org/10.1007/978-3-031-91171-2_6

6.1 Introduction

Safe drinking water is a fundamental need for survival and has been incorporated into the list of Global Sustainable Development goals for 2030 (Bundschuh et al. 2022). Although water covers approximately 70% of the world, only 0.3% of it is accessible for drinking (Jain 2022). The main causes of water pollution and lack of safe drinking water are rapid population growth, industrial operations, technological advancement, and deforestation. Both humans and other living things, as well as marine life, are at risk from this polluted water. Several contaminants such as organic dyes, detergents, pharmaceutical chemicals, pesticides, heavy metals, phenolic compounds, radioactive materials, etc., can affect the water sources and render them unfit for millions of people in both developed and developing nations (Simon et al. 2024; Yadav et al. 2014). These contaminants have the tendency to be poisonous, and their introduction in the biological cycle causes disturbance to both natural and human ecological systems. Out of several heavy metals –Hg, As, Pb, and Cr; arsenic has the most detrimental impact on the neurological, gastrointestinal, renal, and cardiovascular systems due to its toxic nature and lack of biodegradable properties (Jain 2022).

Arsenic (As) is the 12th most commonly encountered and extremely hazardous metalloid with a toxicity rank of 1 and is found in both organic (monomethylarsenic acid and Dimethylarsenic acid) and inorganic forms. There are several sources such as minerals, refineries, agricultural chemicals, and ceramics through which arsenic is released into the surroundings (Elkhatib et al. 2021). Although arsenic is beneficial in many fields, including semiconductor manufacturing, agriculture, but the toxic nature restricts its applications without proper setup for remediation after usage. Four chemical oxidation states of arsenic: Arsenic (As^0), Arsenite (As^{3+}), Arsenate (As^{5+}), and Arsine (As^{3-}) are present in nature. Out of these, As^{3+} and As^{5+} are the predominant found in groundwater and surface water, which is related to the redox potential and pH. Arsenic is one of the heavy metalloids, which mobilizes at pH levels between 6.5 and 8.5 and in both oxidized and reduced environments. The concentration of As^{3+} (H_3AsO_3) is often found in greater amounts than As^{5+} under reduced groundwater conditions and the opposite is also held under oxidized groundwater circumstances. In terms of the toxicity nature among all the different types of Arsenic species, As^{3+} is roughly ~60 times more deadly than As^{5+} because As^{3+} has a strong affinity for sulfurhydral functional groups, which suppress the function of enzymes and potentially damages DNA and RNA molecules at higher intakes, while inorganic forms of Arsenic have roughly having ~100 times more toxicity than organic forms (Overall Toxicity: $As^{3+} > As^{5+} >$ MMA > DMA). There are several cancers (lung, bladder, kidney, liver, skin, etc.) (Group-1 Carcinogen) and non-cancerous diseases (respiratory diseases, anemia, diabetes, weight loss, etc.) that are the consequence of long-term exposure of human being to arsenic environment (Karakoç and Erça 2024). The upper allowed limit concentration of arsenic in the water we drink was lowered from 50 μg/L to 10 μg/L by the World Health Organization (WHO) and the US Environmental Protection Agency (EPA) in 1993 due to its lethal effects (Arsenic 2001).

Arsenic removal is influenced via several parameters including oxidation state, redox potential, and pH, making this extremely complex chemistry for researchers and engineers. Numerous technologies have been developed so far at the lab scale, but their real ground application is still missing or not showing good performance (Rahidul Hassan 2023). Therefore, when developing and implementing any technique for arsenic removal, certain obstacles must be taken into account, such as concentration variation region-wise, presence of other metal ions, adjustment of water pH, long-term operation, and economy. Arsenic is effectively removed from groundwater using several conventional technologies consisting of Ion-exchange, Coagulation, Electrocoagulation, Membrane, and Adsorption (Pezeshki et al. 2023). But every technology has advantages as well as disadvantages, especially when it comes to cost and efficiency which is listed in Table 6.1.

This book chapter highlights the recent advancement in nanotechnology-based solution used for removal of arsenic from polluted water.

6.2 Nanotechnology-Based Solution for Arsenic Removal

Scientists and researchers are exploring the highly promising field of nanotechnology for arsenic removal because of revolutionary advancement in the past few years due to its robust, energy-efficient, affordable, and eco-friendly nature. Furthermore, the current advancement in nanotechnology-based solutions is being driven by mainly two techniques: (i) Absorption-based, and (ii) membrane-based, due to simplicity, lower cost, and efficient technique.

In the upcoming section, nanotechnology-based solutions: Adsorption and membrane based are discussed.

6.2.1 Nanotechnology-Based Solution: Adsorption Based

Adsorption is a surface phenomenon in which molecules are transferred from the bulk to the surface of the adsorbent because of interaction. Traditional adsorbents restrict their high-end application because of several parameters (i.e. surface area, kinetics, active sites, and selectivity). To overcome the challenges, nanomaterials (1–100 nm) are generally utilized for the purpose of adsorption (nanoadsorbents) considering the huge surface area/volume ratio, which enhances the adsorption capability and renders them potentially appropriate for use in applications requiring greater removal efficiency (Bhadouria et al. 2022). Several nanomaterials have recently been explored which can be categorized more specifically as metal/metal oxide-based nano adsorbent (iron, titanium, copper, zinc, manganese, zirconium), carbon-based nanoadsorbent (carbon nanotubes, graphene). The key factors in developing effective nanoadsorbents are (i) Greater surface area and porous structure; (ii) suitable surface functionalization for binding; and (iii) stable and recyclable. The general mechanism

Table 6.1 Different arsenic removal process-advantages, disadvantages, efficiency, and estimated cost

Removal process	Advantages	Disadvantages	Efficiency	Estimated cost	References
Coagulation (Coagulant- $FeCl_3$, Alum, Polyaluminium chloride)	– Utilizing a variety of pH values	– Expensive – Arsenic-contaminated sludge formation – Requires further filtration; – Excessive energy usage	$FeCl_3$: 91% Alum: 96% Polyaluminium chloride: 99.4%	0.32 USD/m^3	(Rajendran et al. 2021; Pezeshki et al. 2023)
Ion exchange	– Rapid process – Minimal toxic sludge formation – Effective technique for getting rid of As^{5+}	– Harmful chemical leaks while regenerating resin – The need for oxidation and/or filtration as pre-and/or post-treatments – Ineffective in getting rid of As^{3+} – Unwanted ions' effect on efficiency – Higher cost	As^{5+}: 95%	0.308 USD/m^3	(Rajendran et al. 2021; Pezeshki et al. 2023)
Electrocoagulation	– Resilient technology – Simple maintenance and repair	– High cost of investment – High energy consumption – Sludge produced – Dependent upon coagulant type and amount	As^{5+}: 99%	–	(Pezeshki et al. 2023)

(continued)

Table 6.1 (continued)

Removal process	Advantages	Disadvantages	Efficiency	Estimated cost	References
Nanotechnology-based solution: Adsorption based	– Minimal operational expense required – Simplicity of use – Low byproduct production – Running without producing sludge – Successful elimination of As^{5+}	– Low efficiency when salts and other organic compounds are present – Need to be treated either before or after (oxidation and/or filtration) – The removal of heavy flocs and the removal of As^{3+} present challenges	– Total As: ~ 100%; 1gm of sulfur-doped copper ferrite-based nanoadsorbent was enough to produce 275 gallon of pure water with no arsenic content and has the capacity to adsorb 181.81 mg of Total As – The removal efficiency reached to 97 ± 4% and 91 ± 3%, for As^{3+} and As^{5+} using MnO_2-modified activated carbon/ granular TiO_2 nanocomposite under real-ground condition – The adsorption capacity for As^{3+} achieved 161.3 mg/g on neutral pH using Zinc ferrite nanoadsorbent under real-time conditions	– –< 0.01 USD/L –	(Duan et al. 2024; Hussain et al. 2024; Shehzad et al. 2024)

(continued)

Table 6.1 (continued)

Removal process	Advantages	Disadvantages	Efficiency	Estimated cost	References
Nanotechnology-based solution: Membrane based	– Easy to use – No produce sludge; – No require chemicals – High elimination effectiveness of arsenate – Minimal energy usage	– High investment costs – Formation of membrane sediment – Impact of undesirable ions on the efficiency	– Nanofiltration: Total As: ~96% – Adsorptive Membrane: >90%	– 0.15 USD/m^3 – Strong correlation with the Nanoadsor-bent cost	(Rajendran et al. 2021; Chia et al. 2023; Pezeshki et al. 2023)

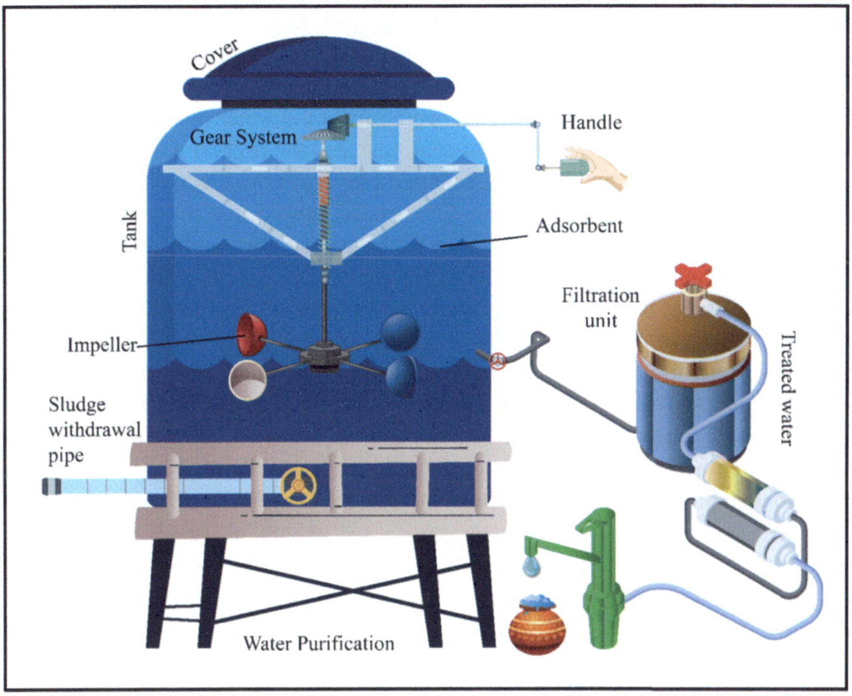

Fig. 6.1 Schematic diagram of the arsenic removal unit using nanoadsorbents (reprinted from Hussain et al. (2024) with permission from Environmental Science: Water Research & Technology. Copyright 2024, The Royal Society of Chemistry)

for the removal of arsenic using nanoadsorbent are complex formation that occurs in three ways: (i) molecules diffuse and cross the boundary layer; (ii) then diffuse across the outside face and enter the adsorbent's pores; (iii) bind to the active site of the interior pores, shown in Fig. 6.6 (a, c) (Ersan et al. 2023). The adsorption unit for arsenic removal using different nanoadsorbents is shown in Fig. 6.1.

In the following sections, various metal/metal oxide-based nanoadsorbents, such as (iron, titanium, copper, zinc, manganese, and zirconium), and carbon-based nanoadsorbents (carbon nanotubes, graphene) are discussed along with their synthesizing technique. In addition, removal efficiency, mechanism, limitations, and other important parameters are highlighted.

6.2.1.1 Iron-Based Nanoadsorbents

Iron-based nanoadsorbents have attracted attention because of their numerous oxidation states, abundant on Earth, and eco-friendly nature. The greatest affinity has been observed for iron oxides. It is well proven that particles of iron oxides and hydroxides that include hematite (α-Fe_2O_3), magnetite (Fe_3O_4), FeOOH, maghemite (γ-Fe_2O_3), and iron oxides combined with different metals and carbon adsorbents show excellent efficiency in removing arsenic due to their strong reactivity, charged surface, stable, cheap cost, and eco-friendly (Elkhatib et al. 2021). Furthermore, Fe_3O_4 and γ-Fe_2O_3 nanoparticles can be easily recovered from the water after adsorption due to their magnetism properties. Prathna et al. revealed the highest adsorption capacities achieved from iron oxide-based nanoparticles are 909 μg/g and 3333 μg/g for both As^{3+} and As^{5+} (TC et al. 2018). The results showed that these nanoparticles had the ability to draw out arsenic from water at a smaller scale setup. To boost arsenic adsorption efficiency, a variety of combinations of iron oxide nanoadsorbents with different metal oxides or supported on different carbon-based plates was studied. The adsorption capacities of various iron oxide-based nanoadsorbents are as follows: Hematite- 9 mg g^{-1} for both As^{3+} and As^{5+}; maghemite- 67 mg g^{-1} and 97 mg g^{-1} for As^{3+} and As^{5+}; Fe-Cu binary oxide- 122 mg g^{-1} and 83 mg g^{-1} for As^{3+} and As^{5+}; aerogel- 216 mg g^{-1} (Ersan et al. 2023). The multimetallic oxide of the iron-based nanoadsorbent has been found to have excellent adsorption capacity (>190 mg/g) in comparison to the other mixed metal oxides based nanoadsorbents (Fig. 6.2 (c)).

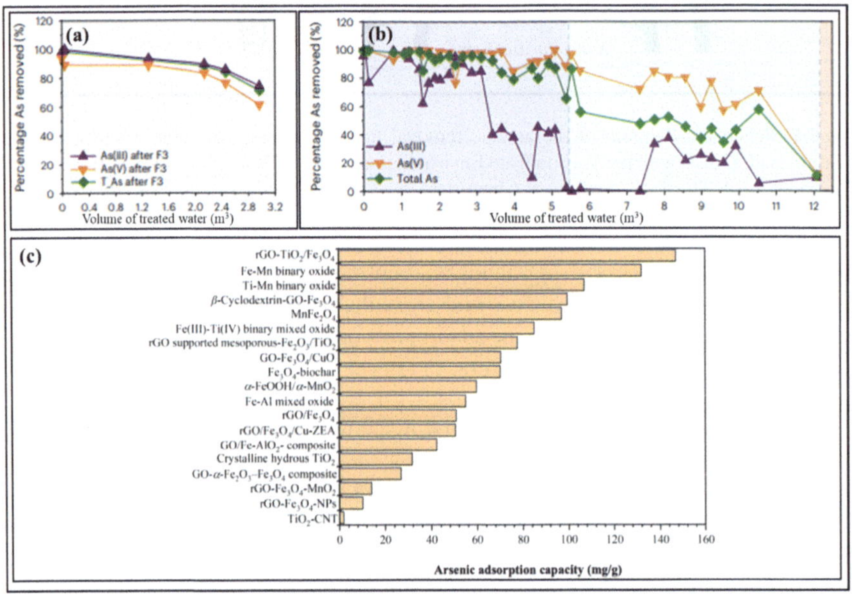

Fig. 6.2 **a, b** Arsenic removal percentages of As^{3+}, As^{5+}, total Arsenic removal (reprinted from Duan et al. (2024) with permission from Nature Water. Copyright 2024, Nature); **c** Adsorption capacities of various metals alone or combined with carbon-based nanoadsorbents for the removal of arsenic (reprinted from Ersan et al., (2023) with permission from Current Opinion in Environmental Science & Health. Copyright 2023, Elsevier B.V.)

6.2.1.2 Titanium-Based Nanoadsorbents

Titanium-based nanoadsorbents (TiO_2@NA) have been highlighted for the arsenic removal application due to their characteristics, such as simple preparation method, low cost, eco-friendly, strong oxidizing capacity, corrosion inhibitor, and showed good adsorption capacity (Bang et al. 2005). Pure TiO_2@NA basically comprises four different kinds of morphologies (i.e., hydrous titania, crystalline, amorphous, and granular) reported for the application of arsenic removal. However, it has been found that amorphous-based TiO2 @ NA adsorption capacities are higher than the crystalline morphology due to the formation of inner surface bidentate complexation for As^{3+} and As^{5+} respectively (Jegadeesan et al. 2010). TiO_2@NA follows the Langmuir adsorption isotherm model for both single as well as multimetallic due to the formation of a monolayer on the surface with no molecular interaction. From comparison studies, TiO_2 bulk particles are estimated to have lower adsorption capabilities than TiO_2@NA (Pena et al. 2006). Ashraf et al., reviewed different homogeneous and heterogeneous TiO_2@NA such as Ce-TiO_2, Zr-TiO_2, Ti-BYC (Basic Yttrium carbonate), and Fe-Ti nanocomposites, among others through which arsenic can be removed (Ashraf et al. 2019). Iron-based nanoadsorbent has been adopted for arsenic removal in wealthier countries since the installation system costs USD 2,740 with an operation cost of USD 1.00 per day on average. Recently, Duan et al., efforted to develop a cost-effective MnO_2-modified activated carbon/granular TiO_2 nanocomposite for arsenic removal under real-world settings that can be used to safeguard the rights of human beings of safe drinking water and to protect the interest of lower-income family. The findings showed the initial arsenic removal efficiency reached $97 \pm 4\%$ and $91 \pm 3\%$, for As^{3+} and As^{5+} which will decline with time but even after 28 months of deployment in the real ground leads to an efficiency of 48% and 57% for As^{3+} and As^{5+} respectively shown in Fig. 6.2 a, b. The cost estimated for the arsenic removal through this route <USD 0.01/L (Duan et al. 2024).

6.2.1.3 Copper-Based Nanoadsorbents

Cupric oxide-based nanoadsorbent is a viable option for arsenic removal from the contaminated water, but the mechanism does not involve the oxidizing process of arsenite to arsenate. With a starting content of arsenic in between 0.1–100 mg/L, the maximum adsorption capacities were observed of 26 and 22 mg g^{-1} over the pH range 6–10 (Martinson and Reddy 2009). Regeneration is one of the important parameters for the adsorbent sustainability aspect and cupric oxide has the potential for solvent regeneration. Hussain et al., developed the sulfur-doped copper ferrite-based nanoadsorbent via a hydrothermal method that can quickly and inexpensively remove the arsenic from groundwater (Hussain et al. 2024). The findings indicated that 275 gallons of safe drinking water could be produced with just one gram of these copper-iron-based nanoadsorbents through the formation of the arsenomolybdate complex. The adsorption mechanism and XPS findings for the sulfur-doped copper ferrite-based nanoadsorbent are shown in Fig. 6.3.

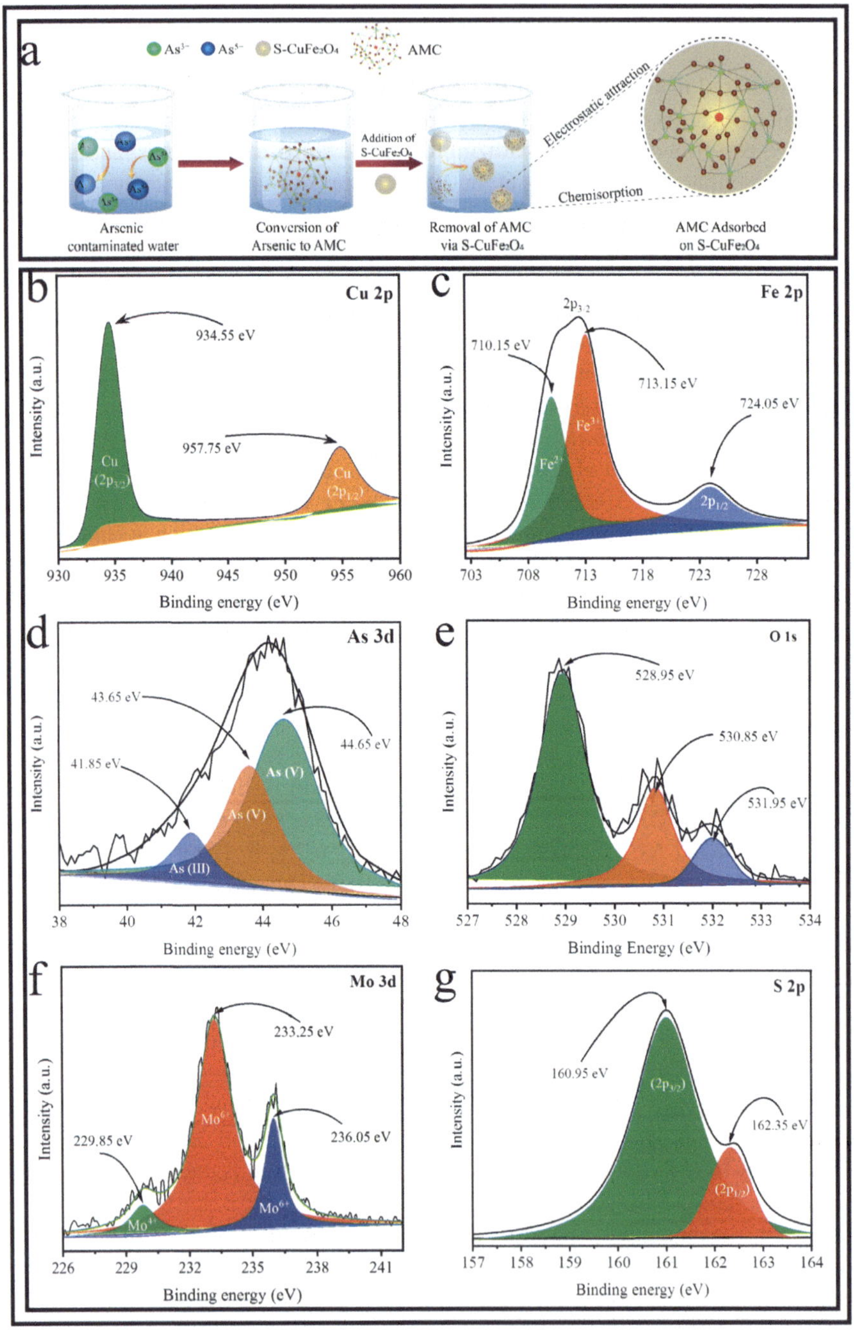

Fig. 6.3 **a** Adsorption mechanism for arsenic removal using sulfur-doped copper ferrite nanostructures; **b**, **c**, **d**, **e**, **f**, **g** XPS analysis using Sulphur doped copper ferrite nanostructures (reprinted from Hussain et al. (2024) with permission from Environmental Science: Water Research & Technology. Copyright 2024, The Royal Society of Chemistry)

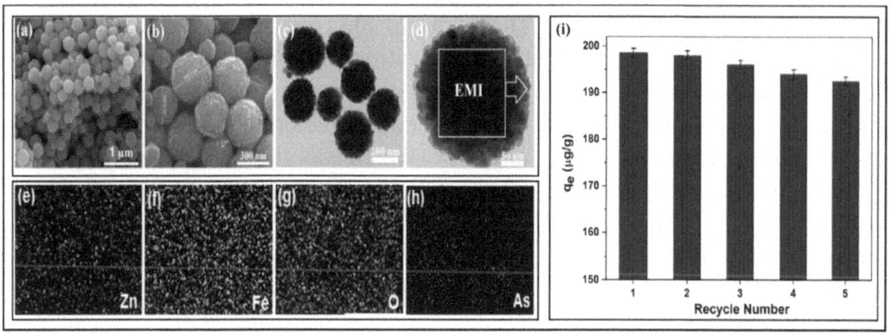

Fig. 6.4 **a**, **b** SEM and **c**, **d** TEM snapshot of Zinc Ferrite nanostructures; **e**, **f**, **g**, **h** mapping of Elemental level images of Zinc ferrite nanostructures with As^{3+}; **i** adsorption capacities of As^{3+} using Zinc ferrite nanostructures for 5 cycles (Reprinted from Shehzad et al. (2024) with permission from Journal of Water Process Engineering. Copyright 2024, Elsevier B.V.)

6.2.1.4 Zinc-Based Nanoadsorbents

Zinc oxide nanoadsorbent (ZnO@NA) is being explored in the field of arsenic removal, which functions well in the pH values between 5.8 and 6.8 but its lower adsorption efficiency of 1.2 mg/g restricts their real-time application. Shehzad et al., studied the zinc ferrite nanoadsorbent prepared by hydrothermal process for the removal of As^{3+}. The results revealed that an adsorption efficiency of 161.3 mg g^{-1} at neutral pH can be achieved due to electrostatic interaction, chemical bond formation, and complex formation. The zinc ferrite nanoadsorbent offers enormous potential, which appears to be a practical solution to remove arsenic (As^{3+}) from polluted water (Shehzad et al. 2024). SEM and TEM images of Zinc Ferrite nanostructures are shown in Fig. 6.4a, b, c, d, demonstrating uniform spherical shape, while Fig. 6.4e, f, g, h displays Zn, Fe, O, As^{3+} are evenly distributed. Figure 6.4 (i) presented the adsorption capacities of As^{3+} using zinc ferrite nanostructures for 5 cycles.

6.2.1.5 Manganese-Based Nanoadsorbents

Manganese oxide-based nanoadsorbent (Mn@NA) has the ability to be an efficient metal oxide adsorbent for removing arsenic because of its complex formation and redox chemistry. Mn@NA can oxidize As^{3+} to As^{5+} for effective arsenic removal, and its adsorption capacity is the function of surface area, crystallinity, pore size, and presence of coexisting ions. Singh et al., developed clumps of δ-MnO_2 nanofiber, and α-MnO_2 nanorods for the removal of arsenic. The morphological study reveals that the 2 × 2 tunnel morphology of α-MnO_2 was made up of needle-shaped nanorods, while the 2D-layered morphology of δ-MnO_2 was made up of small needle fibers grouped that resemble balls. Adsorption experiments revealed that the As^{5+} adsorption capability of 19.41 mg / g in -MnO_2 was strongly dependent on α-MnO_2 and 15.33 mg/g on δ-MnO_2 strongly relied on the pH, α-MnO_2 had a higher rapid equilibrium adsorption capacity than δ-MnO_2. There are two main mechanisms that contribute to As^{5+} adsorption, i.e., ligand exchange and electrostatic force (Singh et al. 2010). In another study, analytical and practical techniques were combined to eliminate arsenic using α-MnO_2 nanofiber. The findings revealed that the optimal adsorption capabilities of 117 and 60 mg/g for As^{3+} and As^{5+}, which was still higher than previously reported Mn@NA in the field of arsenic removal due to stable complex formation in (1 0 0) and (1 1 0) (Luo et al. 2018). Younas et al., reviewed Mn@NA for arsenic removal and highlighted the future challenges and research directions (Younas et al. 2024).

6.2.1.6 Carbon Nanotubes-Based Nanoadsorbents

Carbon Nanotubes (CNTs) have received more attention from researchers and engineers in the field of adsorbents because of their unique characteristics, which include high specific surface area, hollow structure, porous nature, and excellent quick water transport. There are 4 potential active sites that exist in CNTs (i.e., hollow interior, bundle interstitial passage, the convergence points of the parallel tube, external surface). The roll of graphene sheets in cylindrical form results in single-wall CNTs (SWCNTs), whereas several graphene sheets roll together in cylindrical form called Multiwall CNTs (MWCNTs). By combining various metals, the surface of CNTs may be modified to improve their unique characteristics. As a consequence, there are more surface functional groups with better dispersion rates and specific surface areas. To boost the adsorption capacity, CNTs can be altered via oxidation, consolidation with organic molecules, and combination with additional metal ion routes. The iron-based nanocomposite was found to be extremely effective for fast adsorption of As^{5+}, especially for lower initial arsenic concentrations and the concentration increases to affect the removal efficiency to ~ 98.5% at lower pH (Ma et al. 2018). On the other hand, the Zirconia-MWCNT nanohybrid demonstrated excellent adsorption efficiencies of 2 and 5 mg g^{-1} for both As^{3+} and As^{5+}, which is 2–4 times more than the Fe-MWCNT nanocomposite, but the operation time is half for Fe-MWCNTs than Zirconia-MWCNTs (Addo Ntim and Mitra 2012).

6.2.1.7 Graphene-Based Nanoadsorbents

When it comes to removing arsenic from water, graphene-based nanoadsorbents (GNAs) stand out among nanoparticles and nanostructures because of their inherent remarkable adsorption capability, physical and chemical characteristics including high specific surface area, mechanical flexibility, size, shape, and ionic mobility. Graphene is a two-dimensional material consisting of carbon with sp^2 arranged in a hexagonal shape honeycomb structure. Because of its most structured six-membered ring of benzene, it exhibits minimal interactions with the metal ions. The research groups have explored the possibility of functionalization to further improve the physiological properties. The graphene family consists of Graphene oxide (GO), reduced graphene oxide (rGO), and its functionalized and nanocomposite derivatives. GO/rGO is considered as oxidized form of graphene with several oxygen-containing groups (i.e., epoxy groups (C–O–C), hydroxyl groups (–OH) on the basic plane, and carboxyl group (–COOH) at the edges). Furthermore, GO's remarkable adsorption capacity is due in part to both its highly specific surface area and its interaction with the metal ion. DFT studies revealed a binding energy of −4.5 eV between pure GO and arsenic (Panigrahi et al. 2018).

Nevertheless, removal of metal anions is hampered by the negative surface charge of GO/rGO, but their mechanical flexibility and high surface area make them an essential foundation for loading of active nanoparticles in order to increase the adsorption efficiency. Researchers have shown encouraging possibilities to combine iron with graphene-based materials. According to Das et al., the graphene oxide-iron nanohybrid prepared using the sol–gel method through the decoration of Fe-based nanoparticles on the GO base shows a high capacity for quickly adsorbing arsenic, bringing the concentration down to a standard level and effective over the pH range 3–9. The adsorption capacities attained at 20 °C from this GO-based nanohybrid are 306 and 431 mg g^{-1} for As^{3+} and As^{5+}, particularly not much affected by coexisting anions (Das et al. 2020). Pervez et al., conducted studies on the goethite quantum dots impregnated on GO (α-FeOOHQDs@GO) and maximum adsorption efficiencies of 147.38 mg g^{-1} and 69.03 mg g^{-1} for As^{3+} and As^{5+} which is roughly around 4 times higher than β-FeOOH shown in Fig. 6.5i, j (Pervez et al. 2021). TEM images of the α-FeOOHQDs@GO in Fig. 6.5a–d and β-FeOOH@GO in Fig. 6.5e–g show that particles are evenly distributed on the GO surface, whereas SEM image displays the structure of β-FeOOH, featuring a FeOOH nanorod with a spindle shape. Basu et al., studied the hybrid nanocomposite i.e., core–shell microspheres of GO-MnO_2-Goethite for effective elimination of both As^{3+} and As^{5+}, observed the maximal adsorption efficiencies of 27.53 mg g^{-1} and 34.17 mg g^{-1} throughout pH spectrum of 4–6 (Basu et al. 2021).

6.2.2 Nanotechnology-Based Solution: Membrane Based

Researchers have recently shown interest in membrane-based filtration for water purification due to their simplicity, capacity to reduce costs, lack of sludge formation, and ability to recover value-added products. This method is acknowledged as a low-cost, effective way to successfully remove arsenic. The nanofiltration process requires a lot of energy because of the high operating pressure. In recent decades, academics all around the world have given the idea of combining the two currently in use technologies, membrane and adsorption—a lot of thought. Combining these two approaches allows the removal of arsenic with less external force and without affecting the water quality. This can be achieved by incorporating a nanoadsorbent on the microporous membrane. Chia et al., reviewed the current advancement in the field of adsorptive membranes to remove arsenic. The adsorptive membrane is found to be promising for arsenic removal and it has been documented how

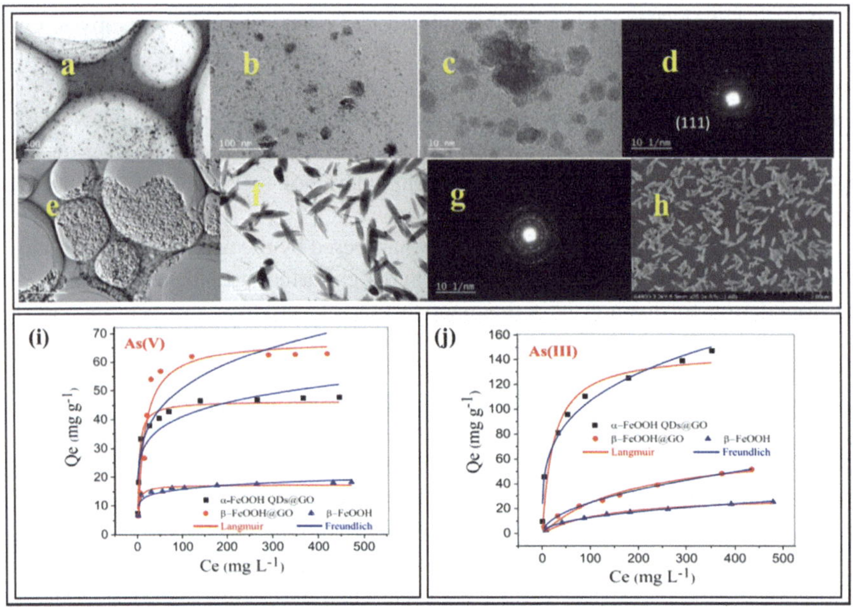

Fig. 6.5 **a-d** TEM snapshot for the α-FeOOHQDs@GO; **e–g** TEM snapshot of the β-FeOOH@GO at various resolution scales; **h** SEM snapshot of the β-FeOOH; **i** Adsorption capacity curve of As^{5+}; **j** Adsorption capacities curve of As^{3+} (reprinted from Pervez et al. (2021) with permission from Science of the Total Environment. Copyright 2021, Elsevier B.V.)

various nanoadsorbents, such as clay, metal oxides, carbon-derived materials, and polymer-incorporated metal–organic frameworks (MOFs) affect the characteristics of membranes composed primarily of polysulfone, polyethersulfone, polyvinylidene fluoride, which will directly impact the arsenic removal efficiency and adsorption capacities. They have listed the challenges associated with the adsorptive membrane: (i) optimum loading of nanoadsorbent without losing membrane properties, (ii) cost of nanoparticles, and (iii) lack of studies on the ground level for a long time (Chia et al. 2023). The adsorption mechanism- physisorption and chemisorption for arsenic removal using an adsorptive membrane displayed in Fig. 6.6b.

6.3 Conclusion and Future Perspectives

The development of technologies that are low-cost is very crucial for providing safe drinking water to millions of people across the globe. In this chapter, recent advancement in the field of arsenic removal using nanotechnology-based solutions have been assessed. Nanotechnology-based solutions are mostly composed of two fields i.e., adsorption and membrane. The majority of the research efforts to date have been directed toward developing new nanoadsorbents having high adsorption

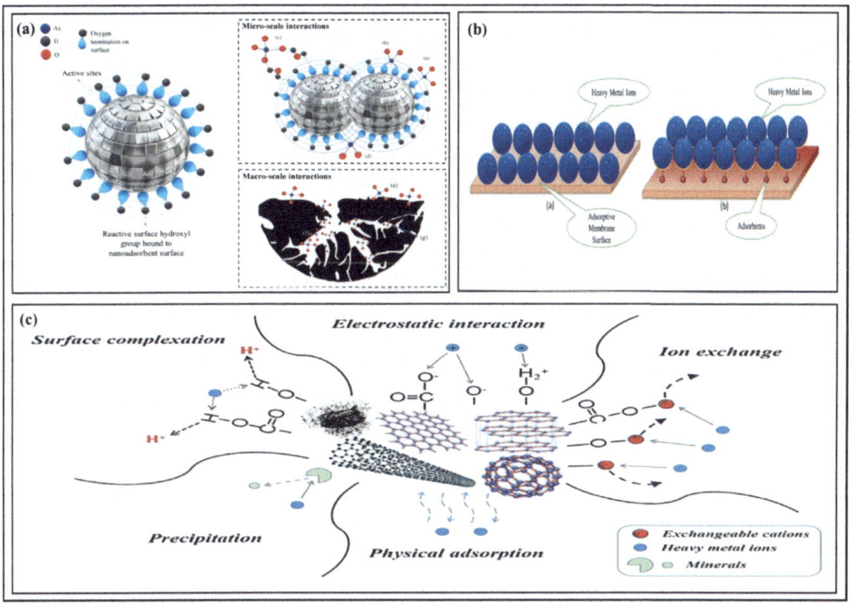

Fig. 6.6 **a** Schematic illustration of micro/macro-scale adsorption mechanism on metal based nanoadsorbents (reprinted from Ersan et al. (2023) with permission from Current Opinion in Environmental Science & Health. Copyright 2023, Elsevier B.V.); **b** Schematic illustration of physiosorption and chemisorption of heavy metal ions in adsorptive membrane nanoadsorbents (reprinted from Chia et al. (2023) with permission from Separation & Purification Reviews. Copyright 2022, Taylor & Francis Group, LLC); **c** Schematic Illustration of process involved in heavy metal adsorption on carbon-based nanoadsorbents (reprinted from Yang et al. (2019) with permission from Chemical Engineering Journal. Copyright 2019, Elsevier B.V.)

efficiency, regeneration capability, cost-effectiveness, eco-friendly, and biocompatible. The adsorption mechanism involves an electrostatic interaction and complex formation with the hydroxyl group terminated. The abundance of metals is one important factor that impacts the price of nanoadsorbent, thereby it is advantageous to choose metals that have substantial reserves and low market cost. Out of the different metal-based nanoadsorbents, iron, and titanium-based nanoadsorbent was extensively investigated by researchers in the decade for the removal of arsenic from groundwater because they were inexpensive, eco-friendly, and easy to collect; although there are issues with cost and agglomeration. A viable approach is to use a metal-based nanoadsorbent supported on carbon-based materials. Activated carbon, CNT, and graphene-based nanoadsorbents have a bright future ahead in the field of arsenic removal because of the larger surface area. The current progress of CNT and graphene has been extensively highlighted in this chapter. Multimetallics supported on cabon-based nanoadsorbent have high potential to remove arsenic because different metals interact synergistically and pass on the benefits of each component to balance the drawbacks. However, the cost of graphene and CNT after functionalization is high, which limits their application despite a high adsorption capacity. Therefore, the focus of future research would be on the regenerability and reusability of CNT and graphene. 3D graphene has the potential for higher arsenic adsorption capacity with a shorter equilibrium time. The development should be taken into account, which will not only reduce the disposal of arsenic treatment but also be economically viable. Hence, there is an urge to develop green technologies for the removal of arsenic and to clean our environment.

The adsorptive membrane is one of the latest nanotechnology-based membranes which is used for the removal of arsenic very effectively and can be easily generated. As a result of the higher cost, there are still limited attempts made to commercialize this membrane. It is a good opportunity to develop a biobased adsorptive membrane that will have a lower cost and perform effectively. It is important to note that there has been very little research in the field of arsenic removal that has been carried out on a real ground setting or commercialization scale. Hence, the future aspect would also be to develop a commercial low-cost setup for arsenic removal.

Acknowledgements The authors sincerely thank Rajiv Gandhi Institute of Petroleum Technology (RGIPT), Jais, India, and other people who are involved in this study for their direct and indirect contributions.

References

Addo Ntim S, Mitra S, 2012. Adsorption of arsenic on multiwall carbon nanotube–zirconia nanohybrid for potential drinking water purification. *J Colloid Interface Sci* 375(1):154–159. https://doi.org/10.1016/j.jcis.2012.01.063

Arsenic WHO, 2001. Environmental health criteria 224. *World Heal Organ Geneva, Switz* :512

Ashraf S, Siddiqa A, Shahida S, Qaisar S, 2019. Titanium-based nanocomposite materials for arsenic removal from water: A review. *Heliyon* 5(5). https://doi.org/10.1016/J.HELIYON.2019.E01577

Bang S, Patel M, Lippincott L, Meng X, 2005. Removal of arsenic from groundwater by granular titanium dioxide adsorbent. *Chemosphere* 60(3):389–397. https://doi.org/10.1016/j.chemosphere.2004.12.008

Basu H, Singh S, Venkatesh M, Pimple MV, Singhal RK, 2021. Graphene oxide-MnO_2-goethite microsphere impregnated alginate: A novel hybrid nanosorbent for As (III) and As (V) removal from groundwater. *J Water Process Eng* 42:102129. https://doi.org/10.1016/j.jwpe.2021.102129

Bhadouria AS, Kumar A, Raj D, Verma A, Singh S, Tripathi P, Kumar Y, Sinha ASK, Tripathi NM, Yeneneh AM, Dwivedi D, 2022. Chapter 8 - Corrosion mitigation in oil reservoirs during CO_2 injection using nanomaterials. In: Sharma T, Chaturvedi KR, Trivedi JJBT-N for CU in OA (eds). *Gulf Professional Publishing*, pp 127–146

Bundschuh J, Niazi NK, Alam MA, Berg M, Herath I, Tomaszewska B, Maity JP, Ok YS, 2022. Global arsenic dilemma and sustainability. *J Hazard Mater* 436:129197. https://doi.org/10.1016/j.jhazmat.2022.129197

Chia RJJ, Lau WJ, Yusof N, Shokravi H, Ismail AF, 2023. Adsorptive Membranes for Arsenic Removal–Principles, Progress and Challenges. *Sep Purif Rev* 52(4):379–399.https://doi.org/10.1080/15422119.2022.2114371

Das TK, Sakthivel TS, Jeyaranjan A, Seal S, Bezbaruah AN, 2020. Ultra-high arsenic adsorption by graphene oxide iron nanohybrid: Removal mechanisms and potential applications. *Chemosphere* 253:126702. https://doi.org/10.1016/j.chemosphere.2020.126702

Duan Y, Sun Y, Palomo A, Li Z, Yang B, Shi Q, Zhang DZ, Yang Q, Meng X, Zheng Y, 2024. MnO_2-modified activated carbon and granular nano-TiO_2 in tandem succeed in treating domestic well water arsenic at point of use. *Nat Water* 2(7):674–683. https://doi.org/10.1038/s44221-024-00268-9

Elkhatib E, Moharem M, Hamadeen H, Mesalem M, 2021. Low-Cost Nanoparticles for Remediation of Arsenic Contaminated Water and Soils. *Arsen Toxic Challenges Solut* :217–251. https://doi.org/10.1007/978-981-33-6068-6_9

Ersan G, Brienza M, Mulchandani A, Apul OG, Garcia-Segura S, 2023. Trends on arsenic species removal by metal-based nanoadsorbents. *Curr Opin Environ Sci Heal* 34:100478. https://doi.org/10.1016/j.coesh.2023.100478

Hussain E, Sultana M, Abid MZ, Buzdar AK, Thabet HK, El-Bahy SM, Jalil M, Rauf A, El-Bahy ZM, Rafiq K, 2024. A report on arsenic removal from water via adsorption of an arsenomolybdate complex on S–$CuFe_2O_4$ adsorbents. *Environ Sci Water Res Technol.* https://doi.org/10.1039/D4EW00300D

Jain R, 2022. Recent advances of magnetite nanomaterials to remove arsenic from water. *RSC Adv* 12(50):32197–32209. https://doi.org/10.1039/D2RA05832D

Jegadeesan G, Al-Abed SR, Sundaram V, Choi H, Scheckel KG, Dionysiou DD, 2010. Arsenic sorption on TiO_2 nanoparticles: Size and crystallinity effects. *Water Res* 44(3):965–973. https://doi.org/10.1016/j.watres.2009.10.047

Karakoç V, Erçağ E, 2024. New Generation Nanoadsorbents and Conventional Techniques for Arsenic Removal from Waters. *J Turkish Chem Soc Sect A Chem* 11(2):845–868. https://doi.org/10.18596/jotcsa.1438869

Luo J, Meng X, Crittenden J, Qu J, Hu C, Liu H, Peng P, 2018. Arsenic adsorption on α-MnO_2 nanofibers and the significance of (100) facet as compared with (110). *Chem Eng J* 331:492–500. https://doi.org/10.1016/j.cej.2017.08.123

Ma M-D, Wu H, Deng Z-Y, Zhao X, 2018. Arsenic removal from water by nanometer iron oxide coated single-wall carbon nanotubes. *J Mol Liq* 259:369–375. https://doi.org/10.1016/j.molliq.2018.03.052

Martinson CA, Reddy KJ, 2009. Adsorption of arsenic(III) and arsenic(V) by cupric oxide nanoparticles. *J Colloid Interface Sci* 336(2):406–411. https://doi.org/10.1016/j.jcis.2009.04.075

Panigrahi P, Dhinakaran AK, Sekar Y, Ahuja R, Hussain T, 2018. Efficient Adsorption Characteristics of Pristine and Silver-Doped Graphene Oxide Towards Contaminants: A Potential Membrane Material for Water Purification. *ChemPhysChem* 19(17):2250–2257. https://doi.org/10.1002/cphc.201800223

Pena M, Meng X, Korfiatis GP, Jing C, 2006. Adsorption mechanism of arsenic on nanocrystalline titanium dioxide. *Environ Sci Technol* 40(4):1257–1262. https://doi.org/10.1021/ES052040E/SUPPL_FILE/ES052040ESI20051215_040329.PDF

Pervez MN, Wei Y, Sun P, Qu G, Naddeo V, Zhao Y, 2021. α-FeOOH quantum dots impregnated graphene oxide hybrids enhanced arsenic adsorption: The mediation role of environmental organic ligands. *Sci Total Environ* 781:146726. https://doi.org/10.1016/j.scitotenv.2021.146726

Pezeshki H, Hashemi M, Rajabi S, 2023. Removal of arsenic as a potentially toxic element from drinking water by filtration: A mini review of nanofiltration and reverse osmosis techniques. *Heliyon* 9(3). https://doi.org/10.1016/j.heliyon.2023.e14246

Rahidul Hassan H, 2023. A review on different arsenic removal techniques used for decontamination of drinking water. *Environ Pollut Bioavailab* 35(1):2165964. https://doi.org/10.1080/26395940.2023.2165964

Rajendran RM, Garg S, Bajpai S, 2021. Economic feasibility of arsenic removal using nanofiltration membrane: A mini review. *Chem Pap* 2021 759 75(9):4431–4444. https://doi.org/10.1007/S11696-021-01694-9

Shehzad K, Haifeng Z, Ahmad M, Yusuf K, Butt MT, Rahmat M, Xu W, Liu J, Xu Y, 2024. Facile synthesis of novel zinc ferrite nanostructures (ZFN) for enhanced adsorption of highly mobile and toxic As(III) from aqueous solutions. *J Water Process Eng* 63:105464. https://doi.org/10.1016/j.jwpe.2024.105464

Simon, M., Joshi, H., Yadav, A. K., & Hopke, P. K. (2024). Simultaneous nitrogen and phosphorus removal from polluted water using a novel bacterial consortium (BACON-38): Promising application for wastewater remediation. Journal of Water Process Engineering, 66, 106044. https://doi.org/10.1016/j.jwpe.2024.106044

Singh M, Thanh DN, Ulbrich P, Strnadová N, Štěpánek F, 2010. Synthesis, characterization and study of arsenate adsorption from aqueous solution by α- and δ-phase manganese dioxide nanoadsorbents. *J Solid State Chem* 183(12):2979–2986. https://doi.org/10.1016/j.jssc.2010.09.023

TC P, K Sharma S, Kennedy M, 2018. Application of Iron Oxide and Iron Oxide/Alumina Nanocomposites for Arsenic and Fluoride Removal: A Comparative Study. *Int J Theor Appl Nanotechnol.* https://doi.org/10.11159/IJTAN.2018.001

Yadav, A. K., Sahoo, S. K., Mahapatra, S., Kumar, A. V., Pandey, G., Lenka, P., & Tripathi, R. M. (2014). Concentrations of uranium in drinking water and cumulative, age-dependent radiation doses in four districts of Uttar Pradesh, India. Toxicological & Environmental Chemistry, 96(2), 192–200. https://doi.org/10.1080/02772248.2014.934247

Yang X, Wan Y, Zheng Y, He F, Yu Z, Huang J, Wang H, Ok YS, Jiang Y, Gao B, 2019. Surface functional groups of carbon-based adsorbents and their roles in the removal of heavy metals from aqueous solutions: A critical review. *Chem Eng J* 366:608–621. https://doi.org/10.1016/j.cej.2019.02.119

Younas M, Bacha AUR, Khan K, Nabi I, Ullah Z, Humayun M, Hou J, 2024. Application of manganese oxide-based materials for arsenic removal: A review. *Sci Total Environ* 918:170269. https://doi.org/10.1016/j.scitotenv.2024.170269

Chapter 7
Adsorptive Materials and Its Techniques for Removal of Arsenic from Wastewater

S. Kavitha, A. Damita Maria Runic, M. Nithishree, M. Meenakshi, D. Dhanuprabha, and A. Arumugam

Abstract Over 200 million people globally are at risk due to increased amounts of arsenic present in drinking water sources in both industrialized and developing nations. The traditional method of removing arsenic from contaminated water is adsorption. Arsenic absorption was traditionally accomplished using materials such as activated charcoal, ion exchange resins, and membrane filtering methods. Although these materials work more effectively, they have several significant disadvantages, including high initial costs, ongoing expenses, and maintenance needs. The implementation of low-cost materials is a pragmatic and cost-effective approach to address this pervasive issue, presenting a strong substitute for costly traditional techniques. Agronomic waste, industrial wastes, natural minerals, and low-cost synthetic adsorbents are among the materials studied. Evaluation of the regeneration potential, kinetics, and adsorption capacity of each material is based on current research. This chapter explores the physicochemical characteristics of surface area, pore size, and functional groups that promote arsenic absorption. It also covers practical concerns such as cost, availability, and environmental effect, and investigates changes to enhance adsorption efficacy. This chapter looks at a critical evaluation of different adsorbents to help choose and optimize low-cost materials for arsenic removal in water treatment, with a focus on scalable and sustainable methods to reduce arsenic pollution.

Keywords Arsenic contamination · Adsorption methods · Low-cost materials · Physicochemical characteristics · Sustainable water treatment

S. Kavitha · A. D. M. Runic · M. Nithishree · M. Meenakshi · D. Dhanuprabha · A. Arumugam (✉)
Advanced Biorefinery & Catalysis Lab, School of Chemical and Biotechnology, SASTRA Deemed University, Thanjavur, Tamil Nadu, India
e-mail: aruchemxl@scbt.sastra.ac.in

A. K. Yadav et al. (eds.), *Arsenic Removal Technologies*,
https://doi.org/10.1007/978-3-031-91171-2_7

7.1 Introduction

Arsenic contamination of water is a critical worldwide environmental problem affecting millions of people globally. Arsenic is a naturally occurring metalloid present in the Earth's crust and has the ability to contaminate water sources through natural and anthropogenic processes. Arsenic in drinking water is a significant health concern, as long-term exposure to arsenic-contaminated water has the ability to induce severe health conditions, such as various forms of cancer, cardiovascular diseases, neurological diseases, as well as developmental diseases. Therefore, the pressing need for efficient arsenic removal technologies is crucial to ensure the integrity of water resources for use in human consumption and to promote environmental sustainability.

Coagulation-flocculation, membrane filtration, precipitation, and ion exchange are some of the traditional arsenic removal techniques from water (Kumar and Chava 2023). However, these traditional techniques are marred with disadvantages of high operating costs, maintenance, and inefficiencies in treating large-scale water contamination problems. The various arsenic removal techniques include adsorption, which has gained widespread attention due to its cost-effectiveness, efficiency, and simplicity. The adsorbent materials are the key elements of arsenic removal from contaminated water by entrapping and immobilizing arsenic ions through surface interactions and preventing their further dispersal. Adsorption techniques for arsenic removal employ various materials ranging from natural minerals, industrial by-products, synthetic adsorbents, and agricultural wastes. Natural minerals like zeolites, kaolinite, and laterite soil have been found to be highly promising in arsenic remediation due to their inherent porosity and high surface reactivity. Similarly, industrial by-products like fly ash and modified rice husk ash have been found to be alternative materials for expensive commercial adsorbents. These materials have twin advantages—solving waste disposal problems while being effective agents for arsenic removal.

Over the past few years, interest in the application of nanomaterials for arsenic adsorption has increased in a very significant manner due to their very high surface area, enhanced reactivity, and good adsorption capacity. Graphene oxide, iron-based nanoparticles, and polymeric ligand exchangers are a few examples of nanocomposites that have been found to be very effective in removing arsenic even at trace concentration levels. Combining nanotechnology with adsorption techniques offers potential pathways for improving arsenic remediation efficiency while remaining within environmental regulations. Agricultural waste materials, such as rice husk, coconut shells, and sawdust, have been extensively explored as low-cost adsorbents for arsenic adsorption. Waste materials are abundant in availability, environmentally friendly, and can be modified to enhance their adsorptive properties. Chemical modification of waste-derived adsorbents, such as metal-oxide activation, has been found to significantly improve their arsenic adsorption capacity. This process not only removes the environmental concerns related to agricultural waste, but also encourages sustainable water treatment processes (Simon et al. 2024).

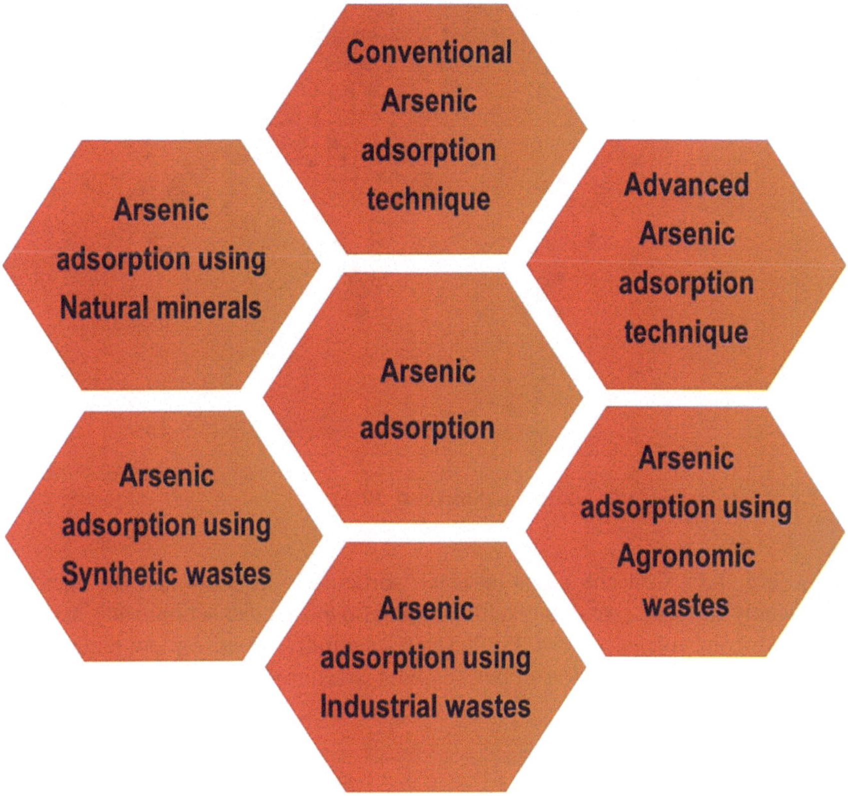

Fig. 7.1 Different techniques used for arsenic adsorption

Synthetic adsorbents, such as metal–organic frameworks (MOFs), ion-exchange resins, and polymer-modified fibers, have also revolutionized arsenic removal techniques in a fundamental way. These materials exhibit high selectivity and regeneration ability and are very efficient in large-scale water treatment processes (Fig. 7.1). The search for innovative adsorbents with better selectivity, stability, and reusability remains a top priority in arsenic remediation technology development (Chowdhury and Mulligan 2011; Shirin et al. 2021).

7.2 Sources of Arsenic Pollution

Arsenic penetrates the soil and groundwater by deteriorating rocks and minerals, followed by leaching and discharge. It may also be discharged into groundwater and soil as a result of human activities. Arsenic is emitted into the atmosphere from sources of high heat, such as burning vegetation, volcanism, and coal-fired power

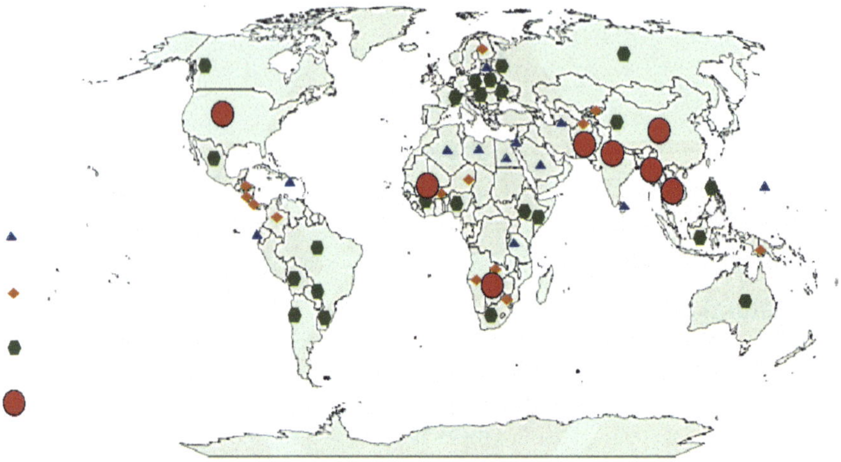

Fig. 7.2. Global arsenic affected regions (Shaji et al. 2021)

plants (Fig. 7.2). Concentrations in water, particularly groundwater, can increase significantly in areas containing sulfide mineral deposits and sedimentary deposits generated from volcanic rocks. Furthermore, naturally occurring low-temperature biomethylation and reduction of arsines emit arsenic into the atmosphere (Shaji et al. 2021).

7.3 Side-Effects of Arsenic Pollution

Long-term exposure to arsenic-tainted water can result in arsenic poisoning, also known as arsenicosis, which can cause skin problems (such as colour changes and rough patches on the palms and soles), skin damage, and cancer. Recent data suggest a link between drinking contaminated water and the development of hypertension, diabetes, and reproductive difficulties. When taken in large quantities, soluble inorganic arsenic can cause acute poisoning, resulting in gastrointestinal distress, brain and circulatory system problems, and death. Survivors may develop disorders such as hemolysis, hepatomegaly, and bone marrow depression. A few studies suggest that drinking water with less than 50 μg of arsenic per litre increases the risk of skin disorders, bladder cancer, and lung cancer.

Arsenic is one of several carcinogens that can cause significant diseases by disrupting the integrity of human cells and their genetic components. Inorganic arsenicals become poisonous when they react with protein sulfhydryl groups or when arsenate replaces and serves as a phosphate group. These inorganic arsenicals have been linked to a variety of health problems, including genotoxicity, skin infections, respiratory, immunological, and endocrine abnormalities, cancer, neurological disorders, liver illness, and gastrointestinal problems. In contrast, inorganic arsenic may represent a dangerous pathway that could promote undesirable biological processes, resulting in skin diseases, cancer, and gastrointestinal problems. Arsenic is one of the many carcinogens that can cause significant diseases by altering the integrity of human cells and genetic components. Inorganic arsenicals are hazardous when they interact with protein sulfhydryl groups or when arsenate replaces and serves as a phosphate group. These inorganic arsenicals have been linked to a variety of health problems, including genotoxicity, skin infections, respiratory, immunological, and endocrine disorders, cancer, neurological concerns, liver illness, and gastrointestinal problems. In contrast, the methylation of inorganic arsenic can be a dangerous pathway that promotes adverse biological processes, possibly leading to skin problems, cancer, and gastrointestinal problems. Early childhood exposure to arsenic, even at low concentrations (less than 50 μg/L), can cause birth abnormalities and negatively affect neurodevelopment, according to empirical studies and experimental findings. Prenatal exposure to arsenic affects the gene expression pathways connected to diabetes, increasing the risk that the disease would develop in maturity (Foroutan et al. 2019).

7.4 Conventional Arsenic Adsorption Technique

Adsorption technologies have proven to be quite effective at purifying water, removing colours, odours, and organic and inorganic pollutants. This is mostly due to the van der Waals separation process and the electrostatic forces of attraction between adsorbed molecules. Arsenic has been removed via ion exchange resins, hydrous metal oxides such as activated alumina, and traditional adsorbents such as activated carbon (AC). New adsorbents, including ferruginous manganese ore, red mud activated by Ce (IV) doping, and porous resin packed with hydrous crystalline zirconium oxide, are being produced and tested (Dutta and Gupta 2022).

7.4.1 Activated Charcoal, Often Known as Activated Carbon (AC)

Due to cost and process stability difficulties, AC is the most commonly utilized adsorbent for a wide range of chemicals. Many agricultural wastes containing carbonaceous minerals, such as lignite, coal, or even coconut shells, could be used to manufacture commercial air conditioners. Activated charcoal is also known as activated carbon or AC. AC is the most commonly used adsorbent for a wide variety of compounds due to cost and process stability issues. Commercial air conditioning could be made from various agricultural wastes containing carbonaceous minerals, such as coal, lignite, and even coconut shells. In this case, AC is used to completely

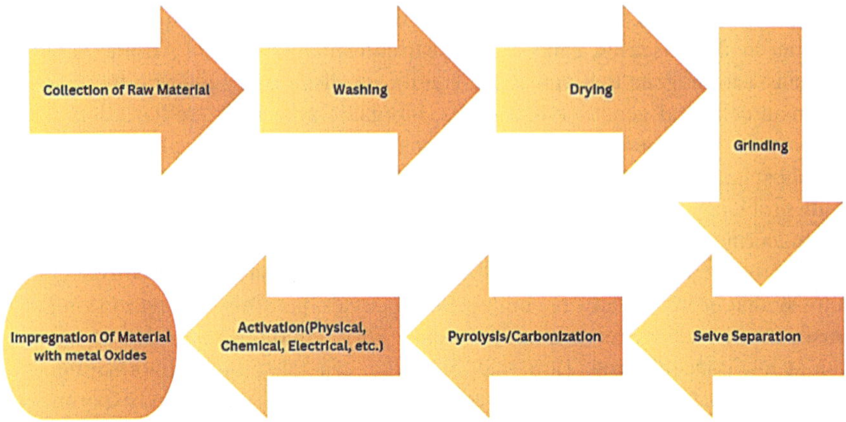

Fig. 7.3 Removal of arsenic by activated charcoal

remove dangerous metals from the aqueous solution because the metal ions and the acidic surface functional groups of AC form surface complexes in.

Figure 7.3. The type of adsorbate, the chemical makeup of AC, the starting pH, the ionic strength, the concentration of the solution, the adsorbent modification techniques, and the physical characteristics all affect removal effectiveness (Chen et al. 2007)

7.4.2 Activated Alumina

Because of their high activity and large surface areas, alumina and modified alumina adsorbents have been widely used to remove arsenic from aqueous solutions. Aluminium hydroxide is hydroxylated to form activated alumina (Al_2O_3). This approach is often used to remove fluoride and arsenic from drinking water. The exact grade, surface area, and pore size of activated alumina (AA), as well as the parameters of the water being treated, all affect AA's ability to absorb arsenic. For field use, an arsenic removal system was built with an activated alumina column, a sand filter, and an EC unit connected in series. Arsenic could be further reduced in concentration by adsorption onto activated alumina. A downstream alumina column can also successfully maintain water quality if the EC system fails or there is intermittent power supply. Sand filters can help activated alumina filters last longer. The field unit provided clean water to the surrounding neighbourhood. During the process, calcium-based deposition was used on the cathodes, and the electrodes were passivated. The proportion of arsenic eliminated by activated alumina was found to range between 55 and 97.5% (Dutta and Gupta 2022).

7.4.3 Ion-Exchange Resins for Arsenic Removal

Ion exchange is a physicochemical treatment that removes arsenic and other contaminants. This approach is widely used in both industrial and municipal water treatment systems because it is extremely effective at treating water with low to moderate arsenic levels. The electrostatically held ions on the surface of the solid phase can be swapped for similarly charged ions from the solution. This successful procedure is commonly used to reduce water hardness and remove pollutants such as chromate anions, arsenate, selenite, and nitrate from contaminated water or wastewater. This method recycles and reinforces exchanged ions by periodically running contaminated water through regenerated resins. Adsorbent materials often contain ions such as hydroxyl (OH^-) and nitrate ($NO_3{}^-$) on their surfaces. When arsenic-containing water passes over the adsorbent in Fig. 7.4, arsenic ions (As^{3+} or As^{5+}) replace surface ions via an ion exchange mechanism. Surface difficulty: Arsenic ions exchange with the adsorbent to create inner-sphere surface complexes. For example, iron oxide adsorbents react with arsenic species to produce Fe-OH groups that then form stable complexes. Treatments such as solid suspension, impurity removal, and filtration are performed before the ion exchange cabin to eliminate organic contaminants that can harm the resins and reduce their effectiveness. Because arsenate has a negative charge, it can be recovered from water more easily than arsenite, which is found mainly as an uncharged element (H_3AsO_3). Arsenite cannot be extracted from water via ion exchange. Total dissolved solids (salt) are one of the variables that reduce the effectiveness of ion exchange in removing arsenic. Hybrid ion exchange processes (HIX-CO_2) remove contaminants from lean brackish water without the use of a semipermeable membrane. Another issue with sludge disposal was caused by the adsorbent regeneration process (Ahmed et al. 2022).

$$R^-[N(CH_3)_3] + Cl^- + H_2AsO_4- \rightarrow R^-[N(CH_3)_3] + H_2AsO_4^- + Cl^-$$

7.4.4 Mineral Adsorbents and Clay for Arsenic Adsorption

Natural clay is an inexpensive, abundant, and readily available resource that could be used in the water and wastewater industries. This study used Iranian natural clay and a clay/Fe–Mn composite to remove harmful arsenic from liquid environments. FTIR, XRD, AFM, EDX, and SEM were used to evaluate the natural clay and clay/Fe–Mn composite. Many variables, including the beginning pH, temperature, sorption dose, and contact time, were studied in relation to the behavior and efficacy of the arsenic(V) adsorption process. The results show that natural clay and clay/Fe–Mn are both effective and recyclable sorbents for removing arsenic(V) from aqueous solutions. Moreover, adding iron and manganese to clay can improve its ability to remove arsenic from it. The initial pH, which controls the speciation of arsenic and

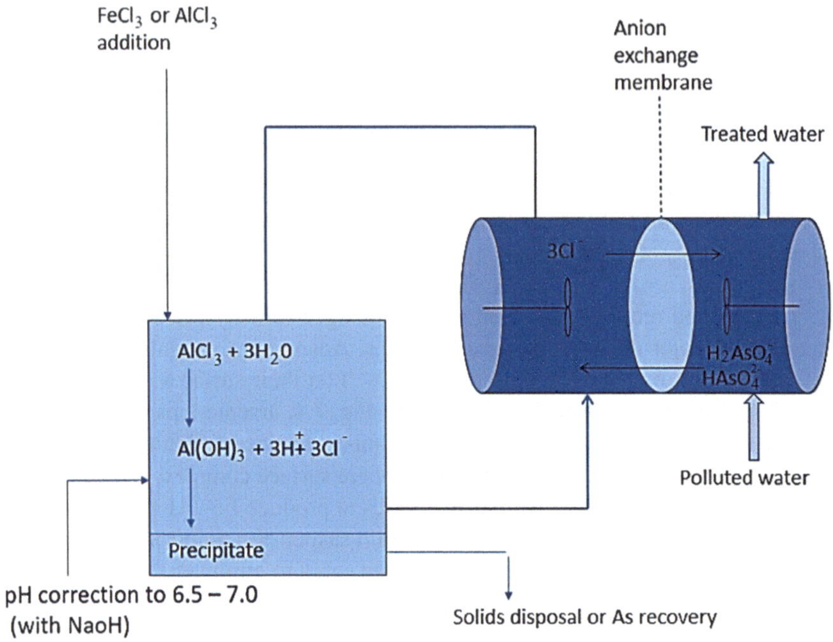

Fig. 7.4 Removal arsenic by ion exchange resins

the surface charge of the adsorbent in aqueous solutions, is one of the important parameters in the adsorption of arsenic(V). Temperature and adsorbent dosage are among the variables (Foroutan et al. 2019).

7.4.5 *Iron-Based Adsorbents*

Ion exchange technology, such as anion exchange resins, was previously thought to be a more successful method of removing arsenic from water. However, it only removes As(V) and is useless against uncharged As(III) species in water. Furthermore, the manufacturing of complicated water filtration systems and ion exchange resin is typically costly. The adsorption capacity was limited due to interference from competitive adsorption of other coexisting anions. The adsorbent regeneration method created a problem with sludge disposal. Reverse osmosis and nanofiltration are two membrane techniques that have recently been extensively reported to be effective at removing arsenic from water. These approaches have the advantages of producing less harmful sludge, being simple to operate, and providing great removal efficiency. However, the initial investment and ongoing costs are relatively substantial; also, a high level of pressure is typically required to force contaminated water through the membranes. Additionally, concentrate discharge, membrane fouling, and flow decrease are unavoidable in the vast majority of membrane operations. Common anions including Cl^-, $NO3^-$, $SO_4{}^{2-}$, and $CO_3{}^{2-}$ did not significantly affect arsenic adsorption for iron-based adsorbents due to the unique chemical interaction between arsenic and iron. According to a report, phosphate can successfully compete with arsenic for adsorption sites, reducing the latter's ability to be adsorbed. Organic debris, such as fulvic acid and humic acid, inhibited arsenic adsorption and delayed the equilibrium of adsorption. Granular activated carbon (GAC) is the most commonly used method for removing organic pollutants from water as a result of its huge specific surface area. Unfortunately, as a result of its negatively charged surface, GAC only adsorbs arsenic species weakly. In recent decades, arsenic adsorption performance has been enhanced by the use of iron-modified activated carbons; in these composite adsorbents, iron oxide particles are the active components that remove arsenic, while GAC provides a vast surface area and serves as a strong support. It has been established that mixing iron oxide nanoparticles and poly(acrylic acid) (GAC) is a feasible strategy to take advantage of both materials' arsenic adsorption properties. The most popular method to generate iron oxide doped GAC is GAC impregnation with an iron salt solution. Lee et al. (2015) used a hydrothermal technique to extract As(V) from water, mixing iron oxides and activated carbon. The Redlich-Peterson model was found to best fit the equilibrium data. The type of iron salt used in the experiment, its pH, duration of treatment, and concentration are all critical factors that influence the arsenic adsorption capacities. Adsorption mechanisms include surface complexation, ion exchange, and electrostatic attraction. The type of iron salt employed in the experiment, its pH, duration of treatment, and concentration are all critical factors that influence the arsenic adsorption capacities. Adsorption mechanisms include surface complexation, ion exchange, and electrostatic attraction. Arsenic adsorption required an iron loading content on GAC of approximately 6% to be the most efficient; curiously, a higher iron content reduced the amount of arsenic that could be adsorbed. At pH values greater than 8.5, phosphate and silicate anions greatly reduced the removal of arsenate, while sulfate, chloride, and fluoride anions had minor impacts (Farhan et al. 2023).

7.5 Advanced Arsenic Adsorption Technique

7.5.1 Polymeric Ligand Exchanger

This was a rising unused innovation recognized as Polymeric Ligand Exchanger (PLE) that was designed by Helfferich (Epstein 1961) (Helfferich 1962). Usually, trapped transition-metal ions like iron or copper are connected to a cross-linked supporting resin to create a PLE. Particle trade in PLEs, in this manner, includes intuitive LAB or metal–ligand complexation due to the nearness of transition metals as terminal utilitarian bunches and electrostatic attraction between the stationary metal particles and the desired anion. Electrostatic intelligent ligands are utilized for the take-up of anion in ordinary ion-exchange tars, but in PLEs, the guideline of ligand quality directs selectivity, permitting more grounded ligands such as arsenate and phosphate for take-up. PLE-produced resins exhibit the following properties: Its take-up of metals ought to be high. To minimize metal leakage during the ligand exchange process, the metal particles need to be strongly bound. To ensure the successful stacking of metal particles in the assembly step, it is essential to leave uncharged useful polymer bundles, otherwise they will not be charged and cannot interact with target ligands. The coordination locales of the metal should not be possessed all by the polymeric gum. The coordination bond could form during the coordination step, in which the metal particles must have ligand interaction. Evacuation by oxidation is necessary for pretreatment within the method on impartial arsenite particles. Ideal for use are biopolymer gels, porous polymers, solid cation-exchange resins, and chelating resins. Biopolymers, such as alginate, may be utilized as cross-linking operators after being immobilized by means of metal ions (Draget et al. 1992). They are categorized based on the kind of metal utilized for impregnation. For occurrence, Fe(III)-Loaded gums and Cu(II)-Loaded gums. Fe(III)-Loaded tars have most of the investigation work done for adsorbing arsenic. Matsugana et al. isolated arsenic from contaminated water with Fe(III)-LDA Resins(lysine-N,N-diacetic) (Matsunaga et al. 1996). The metal-loaded gum had a solid adsorption of arsenate between pHs 2 and 4, and arsenite adsorption was faster at pH 9, but impartial pH makes the PLE wasteful. Fe(III)-filled Moreover, PLE{Fe(III)-ALG} was made from alginate, actually happening biopolymer, and it appeared the most noteworthy capacity for arsenic adsorption between pH values of 3 and 4 (Min and Hering 1998). The type of chelating resin greatly influences the capacity for adsorbing arsenate. Arsenite has a lower adsorption compared to arsenic acid at dilute arsenic concentrations. The average pH controls the adsorption of arsenic. Arsenic acid is adsorbed in strongly acidic mediums, whereas arsenite is adsorbed in alkaline mediums (Mohanty 2017).

Cu(II) loaded resin has a stronger and more specific affinity for arsenate than Fe(III) loaded resin because Cu(II) is a more grounded Lewis corrosive than Fe(III) (Irving and Williams 1949). The Cu(II)-loaded resin was prepared by loading Cu(II) onto a weakly basic chelating resin (DOW-2N). A Cu(II)-based functional group-based mesoporous silica adsorbent was prepared. Compared to commercial SBA, this material showed a greater preference for chromate and arsenate over nitrate and sulfate (Fryxell et al. 1999).

7.5.2 Biological Arsenic Removal

Bacteria are essential in the geochemical cycle of arsenic in groundwater because they participate in oxidation and reduction reactions that impact its properties and mobility. While the neutral arsenite molecule cannot be removed by most technologies, bacteria use arsenate as a terminal electron acceptor. Therefore, oxidation is used as a pretreatment (Mohanty 2017). Bacteria that oxidize iron and manganese such as *Gallionellaferruginea* and *Leptothrixochracea* are considered emerging technology that shows potential for effective elimination of arsenic from underground water sources, as they are very important for the oxidation of arsenic through biological means for treating arsenic-contaminated groundwater (Katsoyiannis et al. 2004). Comparing this new technology with conventional technologies, there are various differences. The process is inexpensive and environmentally friendly, as no chemical reagents are required to oxidize arsenic.

7.5.3 Nanomaterial Arsenic Adsorption

Nanomaterials, characterized by sizes ranging from 1 to 100 nm, possess small dimensions, extensive surface areas, high catalytic properties, and numerous active sites. Nanomaterials are used primarily to absorb toxic and heavy metals, but their tendency to clump together due to their high surface free energy decreases the number of effective adsorption sites (Mohanty 2017). Chandra et al. developed a new method using functionalized graphene in an aqueous medium. They created composites of graphene oxide that exhibit a greater attraction to arsenite and arsenate because of the additional adsorption sites provided by the reduced graphene oxide. This approach is effective for the removal of both arsenite and arsenate from water (Chandra et al. 2010). Zhu et al. created a nanomaterial called graphene with Fe-Fe_2O_3 nanoparticles, which is very efficient in removing arsenite from polluted water. It includes a large number of total adsorption sites, and magnetic separation is possible (Zhu et al. 2012). Luo et al. increased its effectiveness by incorporating MnO_2 nanoparticles into a composite material of Fe_3O_4-RGO (Modified Graphite oxide with Fe_3O_4), expanding the pH range for better arsenic removal by targeting arsenite and arsenate (Luo et al. 2012). Compared to traditional adsorbents nanomaterials have various advantages in arsenic adsorption. These materials have much higher adsorption efficiencies than the conventional adsorbents for both As(III) and As(V) due to their ultra-high surface area.

7.6 Arsenic Adsorption Using Agronomic Waste

Agronomic waste is made up of items such as straw, husks, stems, leaves, and shells that are left over after agricultural operations crop production, harvesting, and processing. These wastes have lately as affordable, environmentally beneficial adsorbents for cleaning up air and water pollutants. This environmentally friendly strategy addresses not only waste management issues but also contamination in the environment. Silica and lignocellulosic chemicals, which are renowned for their porous structure and effective adsorption capacities, are often abundant in agronomic wastes. They can also be transformed into activated carbon, which has a larger surface area, and porous in nature that works well as a sorbent.

Arsenic can usually be immobilized in soil using biochar made from agronomic wastes like sawdust and rice husk, but there is a risk that the soil treated with biochar will allow arsenic to seep into groundwater tables. This problem can be resolved by altering the biochar chemically (using $KMnO_4$ solution) or physically. Additionally, arsenic in the soil can be successfully immobilized through higher temperature pyrolysis (Amin et al. 2020).

Although the process is highly pH dependent, biosorbents, such as rice polish, have shown promise in adsorbing both the oxidative states of arsenic (As(V) and As(III)). The sorption of both types of arsenic using biosorbents, such as rice polish, has been demonstrated to vary dramatically with pH. During a one-hour contact time, pH effect on the sorption of As(V) and As(III) by rice polish was investigated within the initial pH range of 2 to 10. As(III) can be found in two different pH ranges: 2.0 to 9.0 for the nonionic form (H_3AsO_3) and 10 to 12. For anionic form ($H_3AsO_3^-$). According to reports, As(V) is mainly found in the monovalent form (H_2AsO_4) with pH values ranging from 3 to 6. Traces of As(V) may exist as H_3AsO_4 at pH2. But when the pH rises above 8, certain species take center stage (Niazi et al. 2018).

Biochars made from perilla leaves were investigated to determine whether they could be used to remediate groundwater and adsorb As(V) and As(III) from aqueous solutions. Asia's perilla farming produces a large amount of leaf debris, which can be turned into biochar. The study assessed how well perilla leaf biochars, made at high and low temperatures, removed arsenic from water. In batch experiments carried out at pH levels 3–10, biochar produced at 700 °C exhibited the highest As(III) adsorption of upto 88 to 90% at pH 7–9. The initial concentrations of As (V) and As (III) were 0.05–7 mg/L.

With a maximum adsorption efficiency of 11.01 mg/g, the As(III) system is described by the Langmuir adsorption model. Key elements in arsenic sorption have been found, including pore filling, surface complexation, and precipitation (Niazi et al. 2018).

To effectively remove arsenic (As(III)) from water, copper-impregnated activated carbon (CICHC) was made from coconut husk, a naturally occurring material that is widely available. Its performance is assessed in this study in both equilibrium and kinetic scenarios. As(III) was effectively removed from synthetic and industrial wastewater using CICHC; the effectiveness of the removal was controlled by the dose of the sorbent, the duration of contact, and the initial concentration of sorbate. With maximum arsenic adsorption at pH 12, the process is endothermic and adheres to the Langmuir isotherm model. 30% H_2O_2 in a 0.5 M solution can be used to regenerate spent sorbent (Manju et al. 1998). These wastes help with sustainability initiatives and the circular economy due to their quantity, biodegradability, low cost, and general availability.

7.7 Arsenic Adsorption Using Industrial Waste

Research is being done on industrial waste as possible adsorbents to remove heavy metals from wastewater; they have also shown higher adsorption capacity and are a replacement for costly conventional methods.

The protective layer of rice grain made of cellulose is called rice husk. Iron oxide nanoparticles were synthesized and the magnetic modification of rice husk was performed using chemical methods. Iron oxide nanoparticles combined with carbonized rice husk were used to create a highly efficient adsorbent and enhance the conditions of the adsorption process. In adsorption studies to assess the chemical characterization of the adsorbent, EDX-modified and unmodified rice husk ash were compared, and it was found that modified rice husk ash contained higher oxygen, carbon, and silica contents than unmodified rice husk ash. The modified rice husk ash had better adsorption capacity. Their reusability was also investigated; with desirable stability, they are highly efficient and cheap (Javed et al. 2023).

Coal fly ash is a solid waste generated from coal combustion with increased electric power consumption. Fly ash contaminates the environment and wastes a large amount of land. The modified coal fly ash, because of its pore structure, abundant supply, and low cost, is considered a suitable adsorbent. By adding these solid wastes to arsenic solution at room temperature (25 °C) and then adjusting the ph. When the difference between the starting and final concentrations of arsenic in solution was divided by the weight of the adsorbent, the absorption was computed. Arsenic absorption is affected by the pH of the solution. This pH dependent effect is due to the synergistic effects of factors that compete to control the reaction. An increase in the dose of the adsorbent increases arsenic adsorption. It is noteworthy that the utilization of coal fly ash is restricted with a lower specific surface area and arsenic adsorption capacities. This approach helps reduce environmental hazards and realize high-value utilization of coal fly ash (Ochedi et al. 2020).

Activated carbon is produced either naturally or synthetically for water purification. They are natural materials that are implemented for adsorption purposes and also undergo surface modification. There are numerous advantages, such as improved efficiency in removing heavy metals, high porosity, large surface area, and significant reactivity along with irregularity. For this reason, activated carbon is an excellent choice for removing ions from water through adsorption. Even after multiple cycles, the adsorption capacity of the activated carbons for both arsenite and arsenate showed only slight decreases, enhancing reusability. This makes them a suitable adsorbent for arsenic removal (Meez et al. 2021).

Globally, the mining sector generates millions of tons of rock debris, which makes handling environmental and economic problems extremely difficult. Repurposing of Mn/Fe-rich materials for the removal of arsenic and the properties of rock were evaluated. The presence of iron minerals in combination with Al-rich kaolinite resulted in increased arsenic adsorption, while the presence of crystalline or amorphous Mn minerals demonstrated the least amount of adsorption. The presence of bivalent cations enhanced arsenic adsorption. As arsenic was detected in the sample, rock waste retained its natural arsenic adsorbing affinity. Based on the adsorption capacity of these materials, a satisfactory treatment of arsenic-rich groundwater can be conducted (Casentini et al. 2019).

7.8 Arsenic Adsorption Using Synthetic Adsorbents

Because synthetic adsorbents have been very successful in removing arsenic from water, there has been a great deal of interest in using them for this purpose. Different adsorbents have been developed and evaluated for their arsenic removal capabilities, including activated carbon, metal–organic frameworks, and bimetallic oxides.

The Fe–Mn/ATP adsorbent shows strong promise for wastewater is shown by the Fe–Mn/ATP adsorbent, which is made by coprecipitation and is especially effective at removing arsenic ions. Analytical methods that disclose the structural, compositional, and electrochemical properties of the adsorbent include SEM, EDX, XRD, and zeta potential studies. Its cost-effectiveness and reusability are demonstrated by Fe–Mn/ATP, which remarkably retains an arsenic removal efficacy of over 80% even after four regeneration cycles. According to Yu et al. (2024), this bimetallic oxide-based attapulgite-based adsorbent presents a viable and long-term remedy for arsenic pollution in wastewater.

Water sources contaminated with arsenic are a serious environmental problem that can be harmful to human health. One particular substance that has shown great promise as an adsorbent is UiO-66, which can effectively remove arsenic from synthetic wastewater. In comparable tests, its performance outperforms that of other adsorbents including SiO_2 and acid-washed zeolite. It works best when the pH is 9, the temperature is at 25 °C, and the dose is 0.5 mg/ml. These conditions result in sorption efficiency of 10.98 mg/g which is 62.16%. These results demonstrate the potential of UiO-66 to advance water treatment technology, especially in regions impacted by arsenic poisoning, providing a viable means of promoting public health and water quality (Miri 2024).

It has been demonstrated that chemically modified polyacrylonitrile (CM-PAN) fiber works well as an adsorbent to adsorb arsenic from contaminated water bodies. PAN fiber coated with chitosan and iron oxides is used in adsorption operations. Aspen Adsim 11 is used to examine the fiber under different operating conditions for fixed-bed column simulations. Small- and large-scale columns are used in experiments to investigate various bed depths and flow rates. Ion exchange and physisorption are involved in the elimination of arsenic, and the data suit the Langmuir model. Since treated wastewater satisfies home water quality requirements, CM-PAN is a viable option for real-world applications for treating water (Memon et al. 2023).

Magnetized Fe_3O_4 nanoparticles are encased in metal–organic frameworks (MOFs), a new Fe_3O_4/ZIF-8 composite adsorbent was created that is highly effective in adsorbing arsenic As(V) from wastewater. With an adsorption capacity of 116.114 mg/g and a maximum arsenic removal rate of 97.20%, the best results were obtained at a pH of 3 and a dose of 0.60 g/L. Important parameters such as dose, pH, and adsorption kinetics were also assessed. Its stability and structure were validated by characterization techniques. The potential for useful arsenic mitigation in wastewater treatment is shown by thermodynamic analysis, which demonstrates the endothermic nature of the adsorption process (Huang et al. 2022).

The new approach of synthesizing MIL-88B(Fe), a metal organic framework (MOF), in mild circumstances with ethanol as a solvent provides a more environmentally friendly and safe way to remove arsenate from water. Having a 128.99 mg/g high adsorption capacity, MIL-88B (Fe) is very effective at removing arsenic-containing materials. The MOF is granulated with biocompatible chemicals to improve its practical use and facilitate its deployment in water purification systems. Granulated MIL-88B(Fe) has been shown to be a reliable arsenic removal solution that does not result in secondary pollution. It performs better than commercial products in arsenate adsorption and retains long-term efficiency in dynamic column operations (Wu et al. 2022).

7.9 Arsenic Adsorption Using Natural Minerals

Clay minerals can remove both organic and inorganic materials from aqueous environments. Therefore, they have been effectively utilized in arsenic adsorption.

Zeolites are aluminosilicates that fall into the class of alkaline and alkaline earth elements. Due to these favourable chemical and physical properties, for example, silica content, porous structure, and lightweight nature, zeolites are utilized in industrial as well as treatment applications. Sepiolite is a magnesium hydrosilicate mineral, whereas beidellite is a mineral that falls under the category of smectite. These low-cost geomaterials have been proven to successfully remove arsenic ions through identification of the various operating parameters, such as the adsorbent dose, temperature, and pH (Bektas et al. 2011).

Kaolinite is an alumina silicate 1:1 clay mineral group, characterized by its platy structure, particle size, and chemical inertness. They showed outstanding properties when a fluroalylsilane compound was added to the surface of kaolinite. Because they have decent pore-forming ability, they are very popular for the fabrication of reliable microfiltration of inorganic membranes at a reasonable cost. It was found that the adsorption of arsenic in clay minerals is pH dependent. At high pH, kaolinite surfaces become negative, which prevents arsenic adsorption. Thus, these naturally existing substances exhibit maximum arsenic removals and are applied as a significant agent in providing safe consumable water.

Laterite soil is a reddish clay soil whose occurrence has been seen to occur in the tropics and subtropics. Laterite has the potential to remove arsenic as a result of the natural presence of Al (III) and Fe (II). Laterite soil has demonstrated the feasibility of removing arsenic from the aquatic environment. Because they do not need to be chemically pretreated with acids or bases, the expense of preparing the adsorbent is reduced. The advantage of using this as an adsorbent is that iron does not leach into the effluent and the pH of the water remains unchanged during treatment (Maji et al. 2007).

Acknowledgements Arumugam A greatly acknowledges the financial support from DST/SERB, India (CRG/2022/003877).

References

Ahmed, S. F., Kumar, P. S., Rozbu, M. R., Chowdhury, A. T., Nuzhat, S., Rafa, N., Mahlia, T. M. I., Ong, H. C., & Mofijur, M. (2022). Heavy metal toxicity, sources, and remediation techniques for contaminated water and soil. *Environmental Technology and Innovation, 25*, 102114. https://doi.org/10.1016/j.eti.2021.102114

Chowdhury, M. R. I., & Mulligan, C. N. (2011). Biosorption of arsenic from contaminated water by anaerobic biomass. *Journal of Hazardous Materials, 190*(1–3), 486–492. https://doi.org/10.1016/j.jhazmat.2011.03.070

Dutta, N., & Gupta, A. (2022). Development of arsenic removal unit with electrocoagulation and activated alumina sorption: Field trial at rural West Bengal, India. *Journal of Water Process Engineering*, 49(April), 103013. https://doi.org/10.1016/j.jwpe.2022.103013

Farhan, A., Zulfiqar, M., Samiah, Rashid, E. U., Nawaz, S., Iqbal, H. M. N., Jesionowski, T., Bilal, M., & Zdarta, J. (2023). Removal of Toxic Metals from Water by Nanocomposites through Advanced Remediation Processes and Photocatalytic Oxidation. *Current Pollution Reports, 9*(3), 338–358.https://doi.org/10.1007/s40726-023-00253-y

Foroutan, R., Mohammadi, R., Adeleye, A. S., Farjadfard, S., Esvandi, Z., Arfaeinia, H., Sorial, G. A., Ramavandi, B., & Sahebi, S. (2019). Efficient arsenic(V) removal from contaminated water using natural clay and clay composite adsorbents. *Environmental Science and Pollution Research, 26*(29), 29748–29762. https://doi.org/10.1007/s11356-019-06070-5

Fryxell, G. E., Liu, J., Hauser, T. A., Nie, Z., Ferris, K. F., Mattigod, S., Gong, M., & Hallen, R. T. (1999). Design and synthesis of selective mesoporous anion traps. *Chemistry of Materials, 11*(8), 2148–2154. https://doi.org/10.1021/cm990104c

Huang, X., Liu, Y., Wang, X., Zeng, L., Xiao, T., Luo, D., Jiang, J., Zhang, H., Huang, Y., Ye, M., & Huang, L. (2022). Removal of Arsenic from Wastewater by Using Nano Fe3O4/Zinc Organic Frameworks. *International Journal of Environmental Research and Public Health, 19*(17). https://doi.org/10.3390/ijerph191710897

Irving, B. H., & Williams, R. J. P. (1949). *3192 637. 3192.*

Luo, X., Wang, C., Luo, S., Dong, R., Tu, X., & Zeng, G. (2012). Adsorption of As (III) and As (V) from water using magnetite Fe 3O 4-reduced graphite oxide-MnO 2 nanocomposites. *Chemical Engineering Journal, 187*, 45–52. https://doi.org/10.1016/j.cej.2012.01.073

Matsunaga, H., Yokoyama, T., Eldridge, R. J., & Bolto, B. A. (1996). Adsorption characteristics of arsenic(III) and arsenic(V) on iron(III)-loaded chelating resin having lysine-Nα,Nα-diacetic acid moiety. *Reactive and Functional Polymers, 29*(3), 167–174. https://doi.org/10.1016/1381-5148(96)00041-7

Memon, M. suleman, Unar, I. N., & Qureshi, K. (2023). Arsenic Adsorption on Designed Packed Bed Column using Chemically Modified PAN Fiber- A Simulation and Modeling Studies. *International Journal of Applied and Natural Sciences, 1*(1), 51–58. https://doi.org/10.61424/ijans.v1i1.14

Niazi, N. K., Bibi, I., Shahid, M., Ok, Y. S., Burton, E. D., Wang, H., Shaheen, S. M., Rinklebe, J., & Lüttge, A. (2018). Arsenic removal by perilla leaf biochar in aqueous solutions and groundwater: An integrated spectroscopic and microscopic examination. *Environmental Pollution, 232*, 31–41. https://doi.org/10.1016/j.envpol.2017.09.051

Shaji, E., Santosh, M., Sarath, K. V., Prakash, P., Deepchand, V., & Divya, B. V. (2021). Arsenic contamination of groundwater: A global synopsis with focus on the Indian Peninsula. *Geoscience Frontiers*, *12*(3), 101079. https://doi.org/10.1016/j.gsf.2020.08.015

Shirin, S., Jamal, A., Emmanouil, C., & Yadav, A. K. (2021). Assessment of characteristics of acid mine drainage treated with fly ash. Applied Sciences, 11(9), 3910. https://doi.org/10.3390/app11093910

Simon, M., Joshi, H., Yadav, A. K., & Giri, B. S. (2025). Dynamics of pollution and trophic status in selected sub-tropical surface water bodies in Haridwar district, India. Environmental Science and Pollution Research. https://doi.org/10.1007/s11356-024-35724-2

Wu, Y. N., Fang, Y., Fu, J., He, L., Kabtamu, D. M., Matovic, L., Li, F., & Li, J. (2022). Optimized scalable synthesis and granulation of MIL-88B(Fe) for efficient arsenate removal. *Journal of Environmental Chemical Engineering*, *10*(6). https://doi.org/10.1016/j.jece.2022.108556

Zhu, J., Sadu, R., Wei, S., Chen, D. H., Haldolaarachchige, N., Luo, Z., Gomes, J. A., Young, D. P., & Guo, Z. (2012). Magnetic Graphene Nanoplatelet Composites toward Arsenic Removal. *ECS Journal of Solid State Science and Technology*, *1*(1), M1–M5. https://doi.org/10.1149/2.010201jss

Chapter 8
Arsenic Mitigation Technologies from Wastewater Using Low Cost Adsorption Materials

Rajeev Kumar and Jyoti Chawla

Abstract The effects of water contamination have been well demonstrated globally. Water quality is steadily declining as a result of urbanization, industry, and globalization, which is the main cause of health and environmental degradation. Numerous significant amounts of metals & metalloids (Ni, Pb, Cr, Hg, Cd, Cu, Zn, and As) have been found in water samples, raising concerns about water contamination on a worldwide scale. Because arsenic cannot biodegrade, an accumulation of it in biological systems can have harmful impacts on health. According to the US-ATSDR, arsenic ranks first and is a group "A" carcinogen among the hazardous ions list published in 2001. Different methods and materials have been applied to recover arsenic from different immerged water sources. Ion exchange, electrodialysis, membrane separation, chemical oxidation, and adsorption are the prime methods used for recovery of arsenic (As). Adsorption is one of the best methods for arsenic removal because of its low cost and easy handling. Synthetic, natural, and modified materials are effective adsorbents for arsenic removal. The Langmuir, Freundlich, and Temkin isotherms explained the removal mechanism of arsenic using various kinetic models (pseudo-first-order model & pseudo-second-order). The aim of this chapter is to explore the materials that have been used for the mitigation of arsenic from water. This study also compares the materials on the basis of their performance.

Keywords Adsorption · Adsorption mechanism · Arsenic removal · Characterization techniques · Low-cost materials

R. Kumar (✉) · J. Chawla
Department of Applied Sciences, School of Engineering and Technology, Manav Rachna International Institute of Research and Studies, Faridabad, India
e-mail: rajeevkumar.set@mriu.edu.in

A. K. Yadav et al. (eds.), *Arsenic Removal Technologies*,
https://doi.org/10.1007/978-3-031-91171-2_8

8.1 Introduction

Around the world, people are suffering from arsenic poisoning due to drinking contaminated water from various sources. Arsenic-contaminated water has a significant impact on environment and human beings. Arsenic, a naturally occurring element, can be found in all types of water resources due to both natural geological processes and industrial man-made processes, i.e., arsenic-containing pesticides and mining (Nayyar et al. 2022; Shirin et al. 2021). Acute and chronic exposure of arsenic through water system can lead to various severe issues in public health, i.e., cancer, heart disease, kidney failure, and neurological disorder (González-Martnez et al. 2024).

The inorganic form of arsenic is more commonly detected in water, although the organic form is also possible. As arsenite or arsenate, inorganic arsenic can be found. Compared to arsenate, arsenite, the inorganic arsenic, is more deadly and difficult to recover (Gong et al. 2023). More than 230 billion people worldwide and 108 nations suffer from arsenic contaminated water (Shaji et al. 2021). With a higher concentration in Southeast Asian nations, about 180 million people in Asia are in serious danger due to drinking of contaminated water. The groundwater of twenty Indian states and four Indian Union territories was found to be contaminated with arsenic. Groundwater in the Ganga River basin contains more than 4.7 parts per million arsenic (Banerjee et al. 2023).

Numerous researches conducted in the United States, France, China, Belgium, India, Mexico, Bangladesh, and other countries are tracking arsenic in ground and surface water. Research has shown that groundwater contains more arsenic than is allowed and that this poses a serious threat throughout the world. Therefore, the adoption of a cost-effective and practical solution is important for safe drinking water. According to recommendations from several regulatory agencies, the permissible limit of arsenic in drinking water is different for different organization (0.01 mg/l as per World Health Organization, 0.05 mg/l as per Environmental Protection Agency, Indian Council of Medical Research and Indian Standards Institute) (Phat and Le Luu 2023).

After taking these numbers into account, it can be said that arsenic ions in water have gotten to an alarming level in many parts of the world and that efficient and user-friendly techniques are required for mitigation from water. Arsenic is released naturally into water from the weathering of minerals and ores. However, mining, manufacturing processes, such as the electronics industry, and pesticides are the anthropogenic sources of arsenic in water. The removal of desired ions from drinking water resources is a crucial process for ensuring public health. Various methods are employed to remove arsenic from water, with its own pros and cons.

This chapter provides a detailed study of adsorption methodology, the mechanism for arsenic removal, and the characterization of materials. The chapter is structured to achieve the origin and impact of arsenic-contaminated water, and various methods and technologies used to detect and measure arsenic levels in water, providing an in-depth review of the different technologies, methods, and materials for removing arsenic from water using the adsorption method.

8.2 Adsorption Methodology for Arsenic Removal

Purification or treatment of water is a tedious method, that requires many technologies and methodologies for complete purification of water. Waste and ground water contain a lot of toxic chemicals which are required to be removed before this water for further use. In this respect, research has been underway over the last three decades and new methods and technologies have been developed to reduce the cost of the overall process and eco-friendly materials. Biological oxidation, reduction, electrochemical oxidation or reduction, coagulation, precipitation, reverse osmosis, membrane technologies, and adsorption are very effective techniques applied for the treatment of water. Arsenic is one of the most dangerous toxic water contaminants that must be removed before drinking water. In this section, various technologies have been explained to remove all types of contaminants, including arsenic, from water (Fig. 8.1) (Chawla et al. 2015; Saini et al. 2019).

8.2.1 Ion Exchange

It is a reversible technique that can be used to swap out harmful metal ions for safe and environmentally acceptable ones (Kumar et al. 2015a). Because there are no chemical changes made while the ions are exchanging, the process is physical. Arsenic should ideally be in the pentavalent form for removal using anion exchange resin. Ion-exchange resins are highly expensive and require renewal because they clog and lose some of their ability to exchange ions after each cycle. The main drawback of the procedure is the sludge disposal issue that the adsorbent renewal process caused.

8.2.2 Membrane Separation

Also referred to as a pressure-driven process, membrane separation is a commonly employed technique that is highly efficient in separating materials and is easily scaled up. Various membrane types, including ultrafiltration, nanofiltration, and reverse osmosis, have been used (Kumar and Chawla 2013).

Fig. 8.1 Different techniques for removal of arsenic from water

8.2.3 *Oxidation*

Oxidation is the process of oxidizing heavy metal ions found in water by using oxygen from the environment, chemicals, and microbes. Arsenate is created by this procedure from the soluble arsenite. Oxidation is no longer adequate for arsenic cleanup; instead, an adsorption, coagulation, and ion exchange process is required. In addition to atmospheric oxygen, a variety of chemicals and microbes can be used in oxidation. The three most commonly used oxidants in the world's developing world are oxygen, hypochlorite, and permanganate. However, it takes a very long time, whereas permanganate, ozone, and chlorine compounds oxidize quickly. All things considered, this is a difficult procedure that produces an As-bearing residue that is difficult to remove, among other many drawbacks (Singh et al. 2024).

8.2.4 Precipitation

According to Choong et al. (2007) and Fu et al. (2011), precipitation technology consists of using a precipitating agent to turn dissolved contaminants in water into solid precipitates, which are then separated by sedimentation, decantation, and coagulation/flocculation. Water contaminants that are cationic, anionic, or organic can be eliminated by precipitation. The category and amount of the target impurity, the type of precipitant utilized, the presence of coconstituents, the presence of complexing agents, and the pH of the solution all affect the success of a precipitation process that removes impurities. The approach has two drawbacks: A huge amount of sludge is formed, and the precipitating chemicals are expensive (Dhokpande et al. 2024; Shirin et al. 2014).

8.2.5 Electro-Dialysis

Ion-selective membranes are employed in electrodialysis, an electrochemical separation technique that relies on the application of an electrical potential difference. During this process, positive and negative ions are drawn toward their appropriate electrodes by a small DC source. However, this procedure is believed to be ineffective for eliminating arsenic. Using living plants, phytoremediation is a low-cost method of purifying water contaminated (Singh et al. 2024).

8.2.6 Phytoremediation

Phytoremediation is less expensive and more acceptable than older techniques. It is also environmentally favorable, albeit not without restrictions. Due to the slow growth of the plants being studied, the procedure is limited to areas with flora. It also takes a long period (Dhokpande et al. 2024).

8.2.7 Adsorption

Adsorption is the most promising method for adsorbing the contaminants from various water sources. It is a surface activity of the adsorbent and adsorbate. The target material accumulates at the boundaries between two phases in this mass transfer process (Kumar et al. 2023).

8.3 Adsorption Mechanism for Arsenic Removal

It involves the exothermic mass transfer of atoms, molecules, or ions (referred to as the "adsorbate") onto the surface of a substance known as the "adsorbent." Chemisorption and physisorption (chemical adsorption and physical adsorption) are the two main categories of adsorption. Van der Waal's force, for example, may occur between the adsorbed molecules or ions and the adsorbent during physisorption. Although chemisorption is a highly selective process in which monolayer adsorption happens by chemical bonding between the adsorbent and adsorbate material, it is a reversible process (Feisal et al. 2024; Khanzada et al. 2024). In a variety of experimental settings, the target material can be adsorbed from an aqueous environmental solution using selective adsorbents that have high affinity for the pollutant in question. Experimental parameters include the duration of the adsorbent-adsorbate contact, the amount of target ions, the amount of adsorbent used, the pH of the medium, etc. Isotherm/kinetic models have been applied to analyze the experimental data.

8.3.1 Adsorption Isotherms

The isotherms of the process help describe the interaction of adsorbent and ions, providing insight into the adsorption capacity and mechanisms at a constant temperature. Several models are used to describe isotherms, with the most common being Langmuir, Freundlich, and Temkin isotherms (Lin et al. 2020; Siddiqui et al. 2019).

8.3.1.1 Langmuir Isotherm

The Langmuir isotherm is applicable to check the monolayer adsorption of ions homogeneously on the adsorbent. The following equation expressed Langmuir isotherm

$$qe/qmax = KCe/(1 + KCe)$$

In this equation qe represents equilibrium concentration of ions adsorbed (mg/g), qmax represents maximum adsorption capacity (mg/g), K represents Langmuir constant (L/mg), Ce represents equilibrium concentration of ions in solution (mg/L).

8.3.1.2 Freundlich Isotherm

The Freundlich isotherm is applicable to check the multilayer adsorption of ions heterogeneously on adsorbent. The following equation expressed as a Freundlich isotherm

$$\ln qe = \ln K + 1/n \ln Ce$$

K represents Freundlich, 1/n represents heterogeneity factor constant (L/mg).

8.3.1.3 Temkin Isotherm

Temkin isotherm helps to describe the heat of adsorption of the process. The heat of adsorption of the target ions decreases linearly with coverage of the ions on the used adsorbent.

The following equation expressed the Temkin isotherm:

$$qe = B \ln A + B \ln Ce$$

A is constant (L/g), B is calculated from (B = RT/b), b: Temkin constant (J/mol).

8.3.2 Adsorption Kinetics

The kinetics of adsorption describes the rate at which an adsorbate (such as arsenic) is removed from solution by an adsorbent. Understanding the kinetics of arsenic adsorption is crucial to designing efficient water treatment systems. Various models are used to describe the adsorption kinetics (pseudo-first-order (PFO), pseudo-second-order (PSO) and intra-particle diffusion models (IPD)) (Kumar et al. 2022).

8.3.2.1 PFO, PSO, and IDM

Kinetics of removal of ions from the water system is well explained by any one of the PFO, PSO, and IPD models. The following equations explain these three models respectively.

$$\ln(qe - qt) = \ln qe - K_1 t \quad \text{(PFO)}$$

$$t/qt = 1/K_2 qe^2 + t/qe \quad \text{(PSO)}$$

$$qt = K_i t^{0.5} + C \quad \text{(IPD)}$$

where qt at time t amount adsorb, K_1, K_2, and K_i are kinetic constant of PFO, PSO, and IPD respectively, C boundary layer thickness constant.

A graph plotting between ln(qe − qt) vs. t for PFO, t/qt vs. t for PSO, and qt vs. $t^{0.5}$ gives a straight line respectively with − k1, and 1/q slope for PFO and for PSO respectively. C gives boundary layer effect. The high value of the regression coefficient for any of these kinetic models is the best fitted model for adsorption of desired ions from water using an adsorbent.

8.4 Adsorption Materials and Their Characterization

The type of adsorbent, its availability, method of preparation, and its stability play a crucial role before applying any materials as adsorbent for the removal of contaminants from water. Materials are synthesized based on their demand and characteristics properties utilized for targeted contaminates. Characterization of these materials is also essential for verification of the formation of desired materials and presence of essential functional groups on the surface. In this section, both synthetic and natural materials that have been applied as adsorbent for arsenic removal are discussed in detail with their adsorption efficiency.

8.4.1 Materials for Adsorption of Arsenic

Researchers have used a variety of synthetic adsorbent materials over time to remediate arsenic from aqueous systems, and important studies have taken into account pH and the dose into account. Kanel et al. (2005) investigated the utilization of nano iron (zero-valent) in the adsorption of ions (As) (Kanel et al. 2005). Iron hydroxide-coated Al_2O_3 was synthesized by Hlavay et al. (2005), and the reported Freundlich capacity was 3.5 mg/g at 298 K. The Langmuir monolayer efficiency was measured at 3.8 mg/g at a pH of 6.7, dose of 2.5 g/L, duration of 2880 min, and 7.5 mg/L ion concentration (Hlavay and Polyák 2005).

FeO-coated sand was synthesized by Gupta et al. (2005), the modified adsorbent's maximum monolayer efficiency for As(III) ions were 0.028 mg/g at 7.5 pH in 120 min, while the uncoated sand's adsorption capacity was significantly lower (0.0056 mg/g) at the same conditions (V. K. Gupta et al. 2005). 18.2 mg/g of nano (zero valent) Fe may be adsorbed on activated carbon at pH 6.5 (Zhu et al. 2009). Mitigation of As(III) ions activated carbon of Fe-doped micro/nano-particles was reported by Sharma et al. (2010). With 5 g of material at pH 7, the efficiency of monolayer adsorption was 15 mg/g, and the qmax for arsenic ions was determined to be 3.69 mg/g (Sharma et al. 2010).

Multiwall carbon nanotubes and zirconia nano-hybrid were synthesized by Addo Ntim et al. (2012), and at pH 6, their efficiency was reported to be 2 mg/g, requiring 360 min of contact time (Addo Ntim and Mitra 2012). Iron oxide impregnated chitosan beads [(Fe_3O_4)-$Zr(OH)_4$] were prepared by Wang et al. (2014), it was found that 35.3 mg/g arsenic was adsorbed at normal temperature (Wang et al. 2014). δ-MnO2@Fe/Mg- MIL-88B and nanocomposites of hybrid material ZEDA were prepared and used for mitigation of arsenic from water. It was found that δ-MnO2@Fe/Mg- MIL-88B showed 0.1667 mg/g, and ZEDA showed 36 mg/g adsorption capacity (Roy et al. 2013). Copper oxide modified with Boehmite was prepared to mitigate arsenic from water. The Qmax of copper oxide modified with Boehmite was found to be 79.05 mg/g (Panda et al. 2020).

8.4.2 Characterization Techniques for Adsorption of Arsenic

Synthetic and natural materials are effectively applied as adsorbents for mitigation of toxic materials from water. Adsorbent characteristics play a crucial role for adsorption of toxic materials as well as metal ions from water. Adsorbent characteristics can be identified using various characterization techniques. Fourier transform infrared (FTIR), scanning electron microscope (SEM), X-Ray diffraction (XRD), energy dispersive X-ray spectrometry (EDX), atomic emission spectroscopy (AES) are the major characterization techniques used for identification of functional group as well as surface characteristics of adsorbent and concentration of the arsenic in water resources (Fig. 8.2). FTIR provides unique figure print spectrum of various groups & specific functional groups of the adsorbent as well as binding nature of the arsenic on thc surface of the adsorbents (González-Martínez et al. 2024). SEM is an important tool for the identification of the surface characteristics. The adsorption of arsenic on the adsorbent surface depends on the nature of the surface, which may be rough or smooth or containing cavities or pits. The SEM spectrum showed the adsorption of arsenic on the surface by occupying the cavities or pits on the surface (Shaji et al. 2021). XRD is also an effective tool for identification of the nature of the adsorbent, i.e., amorphous or crystalline. Sharp peaks or hump are shown by various XRD spectra of the adsorbents. The maximum number of sharp peaks explained the crystallinity of the adsorbents, while some sharp and some humps represent the intermediate form of the adsorbent. The intensity of the peaks and humps changes after adsorption of arsenic from water on the adsorbents (Nayyar et al. 2022). The abundance of various elements, the composition, and the identification of adsorb arsenic are characterized using EDX technology. The presence of arsenic peaks in the EDX spectrum confirmed the adsorption of arsenic in adsorbent from water resources (Kumar et al. 2022). Toxic materials, heavy metal ions, or arsenic may be present in water resources. AES is a technique for determining the exact concentration of these heavy metal ions, i.e., arsenic in water resources using emission spectrum. The intensity of the emission light determined the concentration of arsenic in the water samples. This technology is very effective and accurate for the determination of arsenic ions.

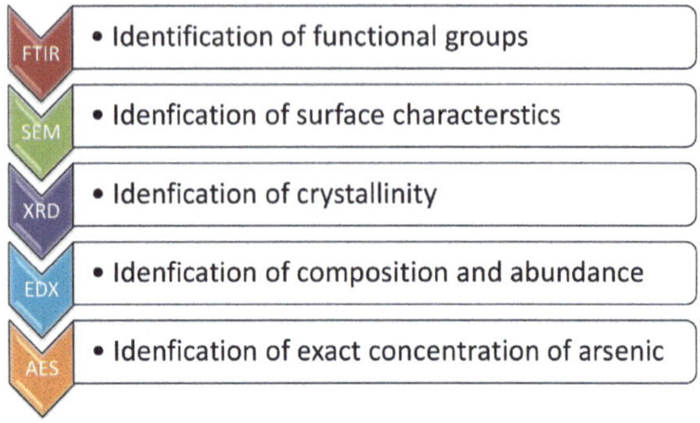

Fig. 8.2 Various characterization techniques for adsorbents and arsenic

8.5 Low-Cost Promising Materials for Arsenic

Synthetic adsorbents have the following primary drawbacks such as their high cost, poor removal effectiveness, and increased environmental hazard. Natural and modified natural materials have also been used to address these issues (Kumar et al. 2015b; Nayyar Kumar et al. 2022a).

The adsorption capacity (in mg/g) of these natural materials also depends on various parameters. Natural materials derived from various parts of plants such waste of fruits, waste of vegetables, plant leaves, roots, stem, fruits part, rice husk, tea waste, bagasse, mango bark, orange peel, rice straw biochar, corn cob, and banana peel are effective materials used as natural occurring adsorbent for mitigation of toxic contaminates and arsenic ions from water. It was observed by various research groups on the basis of various studies that the mitigation efficiency of these natural adsorbents can be increased by addition of some more active functional and high-affinity groups on the surface. In this regard, these natural adsorbents are modified and further used for mitigation of desired ions from water. The thiol group, citric acid, iron, and thioglycolated groups act as active groups for the modification of these natural adsorbents (rice husk biochar, sugar cane bagasse, carbon, watermelon rind) for mitigation of arsenic from water. Natural modified materials are also preferred for the removal of arsenic from water due to their high efficiency (Mohan and Pittman 2007; Shakoor et al. 2016).

Trivalent arsenic remediation from water requires more effective materials and it is also preferable to have an adsorbent that regenerates efficiently up to specific cycles. Numerous investigations have also verified that the nature of solution (acidic or basic) has a significant impact on the sorbate materials' performance and solution chemistry. The slightly acidic to basic nature of the solution has been identified a good condition for removal of arsenic from water.

Various portions of rice plants have been used effectively as adsorbent for removal of toxic contaminates. Rice straw biochar is one of them applied as an adsorbent by Wu et al. (2017) to absorb arsenic ions from water. At pH 10, the metal absorption capacity was observed to be 0.0414 mg/g (Wu et al. 2017). In an experiment, pure momordica charantia (MC) used as adsorbent for arsenic having 0.88 mg/g capacity at basic pH (9.0) and an 5 g/L dose of MC (Pandey et al. 2009).

Activated carbon of various natural occurring materials is highly used as adsorbent for mitigation of toxic contaminates from water. The activated carbon of sugar bagasse has a capacity of 31.25 mg/g in a slightly basic medium (pH = 7.4) at 150 min of contact time in the sample (Tajernia et al. 2014). Egg shell (3.38), watermelon rind (3.4), java plum seed (3.22), pomegranate peel waste (3.10), water chestnut shell (3.04), tea waste (3.0) and corn cob (2.21) mg/g are effectively applied natural adsorbent for arsenic removal with a good adsorption capacity in mg/g at natural pH (Shakoor et al. 2016). Activate carbon of garlic stem (GS) in nanoporous form applied for adsorption of arsenic from water at 6.0 pH. The adsorption capacity of AC of GS was found to be 0.192 (Prajapati and Mondal 2019).

Pentavalent and trivalent ions of arsenic are effectively removed from water using Fe(III) loaded charred sugarcane bagasse. This adsorbent has a high adsorption capacity for petravalent ions (70 m/g) compared to trivalent ions (25 mg/g) of arsenic at slightly basic pH of the solution (8.0) (Chand et al. 2015). Sugarcane bagasse was modified using 2-mercaptoethanol which acts as a more active functional group having a high affinity for arsenic. It showed a high capacity of 28.5 mg/g (Gupta et al. 2015). Biogenic assisted by Pterocarpus mildraedii to kaolin clay (PMKC) by ultrasonic method for removal of petavalent arsenic from water. Slightly acidic (pH = 6) medium was the best medium for 337.22 mg/g of arsenic removal at 40 °C temperature for one hour using 100 mg/L PMKC (Egbosiuba et al. 2024).

Magnetic bleached biochar of tea waste was applied as adsorbent for the removal of arsenic with very high adsorption capacity (714 mg/g). All studies were carried out at room temperature, 9.73 of pH, 0.05 g of adsorbent dose, and 20 mL of 300 mg/L solution of arsenic (McGeogh et al. 2024). Rice husk ash (RHA) and its modified form with FeO nanoparticles were used for effective removal of arsenic from water. A 25.06 mg/g adsorption capacity was achieved using a fabricated adsorbent. Furthermore, this fabricated adsorbent is used as purification cartridge having 98% removal efficiency at 100 mL/min flow rate (Javed et al. 2023). Both mahogany-fruit husk charcoal (MFHC) and banana peduncle charcoal (BPC) were effectively used for the mitigation of arsenic from water with more than 91% and 86% removal at 5.0 pH, and 9.0 pH respectively. Different contact times and adsorbent doses were studies to obtain the optimal conditions for the removal of arsenic (Shaibur et al. 2024).

Phyllanthus emblica seed coat (PESC) and its modified form (MPESC) with thioglycolic acid were applied as an effective adsorbent for the mitigation of arsenic from water. Maximum adsorption capacity (76.92 for PESC and 90.90 mg/g for MPESC) was achieved at pH 7.0. The optimal condition was achieved at 2 h and 0.06 g/L dose of PESC and MPESC (Nayyar et al. 2022). The Prunus armeniaca seed coat (PASC) and its modified form (MPASC) with thioglycolic acid were applied as effective adsorbent for mitigation of arsenic from water. Maximum adsorption capacity (125 for PASC and 142.85 mg/g for MPESC) was achieved at pH 7.0. Optimum condition was achieved at 2 h and a 0.06 g/L dose of PASC and MPASC (Nayyar et al. 2022). The punica granatum carpellary membrane was used for the mitigation of arsenic from water as a natural adsorbent. Maximum adsorption capacity (83.33 mg/

g) was achieved at pH 5.0. Optimum condition was achieved at 2 h and 0.06 g / L of adsorbent (Nayyar et al. 2022).

8.6 Conclusions

Both synthetic and natural adsorbents have been effectively used for arsenic mitigation over water resources from the past decade. Recently, more emphasis has been placed on eco-friendly and cost-effective materials for arsenic removal. Waste of fruits and vegetables is an effective adsorbent used for arsenic mitigations. The adsorption capacity of the natural adsorbents was improved by the addition of more active and high affinity groups on the surface of the adsorbent. Egg shell, watermelon rind, java plum seed, pomegranate peel waste, water chestnut shell, tea waste, and corn cob and some modified adsorbent, i.e., sugarcane bagasse, garlic stem, and many more are very effective adsorbent for mitigation of arsenic from water at natural pH. The pH of the solution, the time of contact, the dose of materials, and the concentration of the solution are crucial factors that affected the rate of adsorption of targeted ions from the water. The solid–liquid system that causes the remediation of arsenic ions from water and causes the accumulation of arsenic at the solid interface is pertinent to the current investigation.

Acknowledgements The authors sincerely thank our prestigious institute MRIIRS, SET, for providing support for writing of this chapter.

References

Addo Ntim, S., & Mitra, S. (2012). Adsorption of Arsenic on Multiwall Carbon Nanotube–Zirconia Nanohybrid for Potential Drinking Water Purification. *Journal of Colloid and Interface Science, 375*(1), 154.https://doi.org/10.1016/J.JCIS.2012.01.063

Banerjee, S., Dhar, S., Sudarshan, M., Chakraborty, A., Bhattacharjee, S., & Bhattacharjee, P. (2023). Investigating the synergistic role of heavy metals in Arsenic-induced skin lesions in West Bengal, India. *Journal of Trace Elements in Medicine and Biology : Organ of the Society for Minerals and Trace Elements (GMS), 75.* https://doi.org/10.1016/J.JTEMB.2022.127103

Chand, T. H., Raj, P. M., Nath, G. K., & Deba Bahadur, K. (2015). Removal of Arsenic from Aqueous Solution Using Iron(III)-loaded Sugarcane Bagasse. *Research Journal of Chemical Sciences, 5*(11), 51–58. www.isca.me

Chawla, J., Kumar, R., & Kaur, I. (2015). Carbon nanotubes and graphenes as adsorbents for adsorption of lead ions from water: a review. *Journal of Water Supply: Research and Technology-Aqua, 64*(6), 641–659. https://doi.org/10.2166/AQUA.2015.102

Dhokpande, S. R., Deshmukh, S. M., Khandekar, A., & Sankhe, A. (2024). A review outlook on methods for removal of heavy metal ions from wastewater. *Separation and Purification Technology, 350*, 127868. https://doi.org/10.1016/J.SEPPUR.2024.127868

Egbosiuba, T. C., Tran, T. Q., Arole, K., Zhang, Y., Enyoh, C. E., Mustapha, S., Tijani, J. O., Yadav, V. K., Anadebe, V. C., & Abdulkareem, A. S. (2024). Biotreatment of clay-based adsorbent to eliminate arsenic (V) ions and malachite green from wastewater: Isotherm, kinetics, thermodynamics, reusability and mechanism. *Results in Engineering*, *22*, 102073. https://doi.org/10.1016/J.RINENG.2024.102073

Feisal, N. A. S., Kamaludin, N. H., Ahmad, M. A., & Tengku Ibrahim, T. N. B. (2024). A comprehensive review of nanomaterials for efficient heavy metal ions removal in water treatment. *Journal of Water Process Engineering*, *64*, 105566. https://doi.org/10.1016/J.JWPE.2024.105566

Gong, S., Yang, J., Zhou, W., Liu, X., & Wang, D. (2023). Fe-containing materials for inorganic/organic arsenic removal from water: A review of current status and future prospects. *Journal of Cleaner Production*, *428*, 139533. https://doi.org/10.1016/J.JCLEPRO.2023.139533

González-Martínez, F., Johnson-Restrepo, B., & Quiñones, L. A. (2024). Arsenic inorganic exposure, metabolism, genetic biomarkers and its impact on human health: A mini-review. *Toxicology Letters*, *398*, 105–117. https://doi.org/10.1016/J.TOXLET.2024.06.008

Gupta, V. K., Saini, V. K., & Jain, N. (2005). Adsorption of As(III) from aqueous solutions by iron oxide-coated sand. *Journal of Colloid and Interface Science*, *288*(1), 55–60. https://doi.org/10.1016/J.JCIS.2005.02.054

Gupta, A., Vidyarthi, S. R., & Sankararamakrishnan, N. (2015). Concurrent removal of As(III) and As(V) using green low cost functionalized biosorbent – Saccharum officinarum bagasse. *Journal of Environmental Chemical Engineering*, *3*(1), 113–121. https://doi.org/10.1016/J.JECE.2014.11.023

Hlavay, J., & Polyák, K. (2005). Determination of surface properties of iron hydroxide-coated alumina adsorbent prepared for removal of arsenic from drinking water. *Journal of Colloid and Interface Science*, *284*(1), 71–77. https://doi.org/10.1016/J.JCIS.2004.10.032

Javed, K., Mahmood, S., Ammar, M., Abbas, N., Shah, M. Y., Ahmed, T., & Mustafa, G. (2023). Rice husk ash adsorbent modified by iron oxide with excellent adsorption capacity for arsenic removal from water. *International Journal of Environmental Science and Technology*, *20*(3), 2819–2828. https://doi.org/10.1007/S13762-022-04390-7/METRICS

Kanel, S. R., Manning, B., Charlet, L., & Choi, H. (2005). Removal of Arsenic(III) from Groundwater by Nanoscale Zero-Valent Iron. *Environmental Science and Technology*, *39*(5), 1291–1298. https://doi.org/10.1021/ES048991U

Khanzada, A. K., Al-Hazmi, H. E., Kurniawan, T. A., Majtacz, J., Piechota, G., Kumar, G., Ezzati, P., Saeb, M. R., Rabiee, N., Karimi-Maleh, H., Lima, E. C., & Mąkinia, J. (2024). Hydrochar as a bio-based adsorbent for heavy metals removal: A review of production processes, adsorption mechanisms, kinetic models, regeneration and reusability. *Science of The Total Environment*, *945*, 173972. https://doi.org/10.1016/J.SCITOTENV.2024.173972

Lin, S., Jin, J., Sun, S., & Yu, J. (2020). Removal of arsenic contaminants using a novel porous nanoadsorbent with superior magnetic recovery. *Chemical Engineering Science: X*, *8*, 100069. https://doi.org/10.1016/J.CESX.2020.100069

McGeogh, M., Annath, H., & Mangwandi, C. (2024). Turning teawaste particles into magnetic bio-sorbents particles for arsenic removal from wastewater: Isotherm and kinetic studies. *Particuology*, *87*, 179–193. https://doi.org/10.1016/J.PARTIC.2023.08.003

Mohan, D., & Pittman, C. U. (2007). Arsenic removal from water/wastewater using adsorbents—A critical review. *Journal of Hazardous Materials*, *142*(1–2), 1–53. https://doi.org/10.1016/J.JHAZMAT.2007.01.006

Nayyar, M., Chawla, J., & Kumar, R. (2022). N-Substituted 2-Sulfanylacetamide Prunus armeniaca: Synthesis, Characterization, and Adsorption Studies for As(III) Remediation. *Russian Journal of Applied Chemistry*, *95*(5), 742–755. https://doi.org/10.1134/S1070427222050159/METRICS

Panda, A. P., Jha, U., & Swain, S. K. (2020). Synthesis of nanostructured copper oxide loaded boehmite (CuO_Boehmite) for adsorptive removal of As(III/V) from aqueous solution. *Journal of Water Process Engineering*, *37*, 101506. https://doi.org/10.1016/J.JWPE.2020.101506

Pandey, P. K., Choubey, S., Verma, Y., Pandey, M., & Chandrashekhar, K. (2009). Biosorptive removal of arsenic from drinking water. *Bioresource Technology*, *100*(2), 634–637. https://doi.org/10.1016/J.BIORTECH.2008.07.063

Phat, V. V., & Le Luu, T. (2023). Review on the removal of Arsenic in groundwater using laterite adsorbent. *Current Opinion in Environmental Science & Health, 35*, 100496. https://doi.org/10.1016/J.COESH.2023.100496

Prajapati, A. K., & Mondal, M. K. (2019). Hazardous As(III) removal using nanoporous activated carbon of waste garlic stem as adsorbent: Kinetic and mass transfer mechanisms. *Korean Journal of Chemical Engineering, 36*(11), 1900–1914. https://doi.org/10.1007/S11814-019-0376-X/METRICS

Roy, P., Mondal, N. K., Bhattacharya, S., Das, B., & Das, K. (2013). Removal of arsenic(III) and arsenic(V) on chemically modified low-cost adsorbent: Batch and column operations. *Applied Water Science, 3*(1), 293–309. https://doi.org/10.1007/S13201-013-0082-5/TABLES/3

Saini, S., Kumar, R., Chawla, J., & Kaur, I. (2019). Adsorption of bivalent lead ions from an aqueous phase system: Equilibrium, thermodynamic, kinetics, and optimization studies. *Water Environment Research, 91*(12), 1692–1704. https://doi.org/10.1002/WER.1180

Shaibur, M. R., Khatun, Y., Howlader, M., Islam, M. M., Rahman, M. W., Khan, A. S., & Ambade, B. (2024). Determination of water quality and efficient removal of arsenic and iron from groundwater using mahogany fruit husk and banana peduncle charcoals. *Results in Engineering, 22*, 102220. https://doi.org/10.1016/J.RINENG.2024.102220

Shaji, E., Santosh, M., Sarath, K. V., Prakash, P., Deepchand, V., & Divya, B. V. (2021). Arsenic contamination of groundwater: A global synopsis with focus on the Indian Peninsula. *Geoscience Frontiers, 12*(3), 101079. https://doi.org/10.1016/J.GSF.2020.08.015

Shakoor, M. B., Niazi, N. K., Bibi, I., Murtaza, G., Kunhikrishnan, A., Seshadri, B., Shahid, M., Ali, S., Bolan, N. S., Ok, Y. S., Abid, M., & Ali, F. (2016). Remediation of arsenic-contaminated water using agricultural wastes as biosorbents. *Critical Reviews in Environmental Science and Technology, 46*(5), 467–499. https://doi.org/10.1080/10643389.2015.1109910

Sharma, A., Verma, N., Sharma, A., Deva, D., & Sankararamakrishnan, N. (2010). Iron doped phenolic resin based activated carbon micro and nanoparticles by milling: Synthesis, characterization and application in arsenic removal. *Chemical Engineering Science, 65*(11), 3591–3601. https://doi.org/10.1016/J.CES.2010.02.052

Shirin, S., Jamal, A., Emmanouil, C., & Yadav, A. K. (2021). Assessment of characteristics of acid mine drainage treated with fly ash. Applied Sciences, 11(9), 3910. https://doi.org/10.3390/app11093910

Shirin, S., & Yadav, A. K. (2014). Physico-chemical analysis of municipal wastewater discharge in Ganga river, Haridwar district of Uttarakhand, India. Current World Environment, 9(2), 536–543. https://doi.org/10.12944/CWE.9.2.39

Siddiqui, S. I., Naushad, M., & Chaudhry, S. A. (2019). Promising prospects of nanomaterials for arsenic water remediation: A comprehensive review. *Process Safety and Environmental Protection, 126*, 60–97. https://doi.org/10.1016/J.PSEP.2019.03.037

Singh, A., Shah, S. S., Sharma, C., Gupta, V., Sundramoorthy, A. K., Kumar, P., & Arya, S. (2024). An approach towards different techniques for detection of heavy metal ions and their removal from waste water. *Journal of Environmental Chemical Engineering, 12*(3), 113032. https://doi.org/10.1016/J.JECE.2024.113032

Tajernia, H., Ebadi, T., Nasernejad, B., & Ghafori, M. (2014). Arsenic removal from water by sugarcane bagasse: An application of response surface methodology (RSM). *Water, Air, and Soil Pollution, 225*(7), 1–22. https://doi.org/10.1007/S11270-014-2028-4/METRICS

Wang, J., Xu, W., Chen, L., Huang, X., & Liu, J. (2014). Preparation and evaluation of magnetic nanoparticles impregnated chitosan beads for arsenic removal from water. *Chemical Engineering Journal, 251*, 25–34. https://doi.org/10.1016/J.CEJ.2014.04.061

Wu, C., Huang, L., Xue, S. G., Huang, Y. Y., Hartley, W., Cui, M. qian, & Wong, M. H. (2017). Arsenic sorption by red mud-modified biochar produced from rice straw. *Environmental Science and Pollution Research International, 24*(22), 18168–18178.https://doi.org/10.1007/S11356-017-9466-7

Zhu, H., Jia, Y., Wu, X., & Wang, H. (2009). Removal of arsenic from water by supported nano zero-valent iron on activated carbon. *Journal of Hazardous Materials*, *172*(2–3), 1591–1596. https://doi.org/10.1016/J.JHAZMAT.2009.08.031

Chapter 9
Sustainable Low-Cost Adsorbents for Arsenic Removal: Advances, Challenges, and Future Prospects

Blessy Thankachan, Thomas J. Kallupurakel, Neethu Shajan, and Vishnu Manirethan

Abstract Introduced into the environment through both natural processes and anthropogenically, arsenic (As) is a toxic metalloid that causes major maladies including cancer. Declared a Group I human carcinogen by the IARC, the WHO has established a 10 μg/L limit for inorganic arsenic in drinking water. Concentrations beyond this level have, however, been found in more than 50 nations, endangering over 200 million people in notably economically impoverished areas. Currently deployed arsenic removal techniques such as chemical precipitation and ion exchange rely on infrastructure and are prohibitively costly. Low-cost, economically viable adsorbents from minerals, industrial waste, and biomass are under investigation as a sustainable substitute for their efficacy. Recent developments improve these materials by means of metal-doped oxides, nanocomposites, and bio-based additives. Promising low-cost adsorbents, their adsorption capacity, and future use for reducing arsenic contamination are investigated in this chapter.

Keywords Arsenic adsorption · Arsenic contamination · Low-cost adsorbents · Heavy metal remediation

9.1 Introduction

Historically dubbed 'The Silent Killer' and 'Inheritance Powder,' the infamous metalloid arsenic has long been associated with covert poisonings and untimely deaths throughout history. Since arsenic has no taste or smell, it was frequently used as a poison by mixing it with meals and beverages (Kuivenhoven and Mason 2023).

B. Thankachan · T. J. Kallupurakel · V. Manirethan (✉)
Department of Chemical Engineering, Amal Jyothi College of Engineering, Kanjirappally, India
e-mail: vishnumvishnu@gmail.com; mvishnu@amaljyothi.ac.in

N. Shajan
Department of Chemical and Bioprocess Engineering, Hamburg University of Technology, Hamburg, Germany

A. K. Yadav et al. (eds.), *Arsenic Removal Technologies*,
https://doi.org/10.1007/978-3-031-91171-2_9

Arsenic, a noted carcinogen, is classified as a Group I human carcinogen by the International Agency for Research on Cancer, and chronic exposure to it through contaminated groundwater can pose serious health risks, including kidney cancer, skin lesions, neurological disorders, and liver damage, with mortality occurring in extreme cases (Rehman et al. 2020). The toxicity of arsenic contamination in groundwater mirrors the challenges posed by other persistent pollutants like fluoride, which also threatens human health and requires advanced remediation technologies (Kumar et al. 2023).

Arsenic exists in both organic and inorganic forms, with the latter being more toxic and is introduced into the environment through both anthropogenic and natural means. Among its inorganic species, arsenite (As(III)) is more hazardous due to its higher bioavailability and strong affinity for sulfhydryl (–SH) groups, which disrupt enzymatic functions, making it harder to remove from water. Arsenic speciation depends primarily on redox conditions. In oxidizing environments like surface water, arsenic is found as arsenate (As(V)), while reducing conditions in groundwater favor arsenite (As(III)). Though As(III) is more toxic, arsenate contamination is widespread, requiring urgent remediation. pH also influences arsenic speciation. Under oxidizing conditions, arsenic appears as arsenic acid (H_3AsO_4) at $pH < 2$ and as $H_2AsO_4^-/HAsO_4^{2-}$ at pH 2–11. In reducing environments, arsenous acid dominates, transitioning to $HAsO_3^{2-}$ at $pH > 12$. These variations complicate removal strategies, emphasizing the need for efficient, cost-effective arsenic treatment solutions (Chutia et al. 2009; Kumarathilaka et al. 2018).

With the advent of modern technology, modelling technologies, and survey methods, a number of studies from a myriad of regions have determined groundwater contamination by toxic pollutants such as arsenic, is more extensive than commonly realized and extends from under-developed to developed countries, including Argentina, Bangladesh, Canada, China, Hungary, India, Japan, Mexico, Nepal, Taiwan, and the USA (Adeloju et al. 2021). Fluoride contamination, often found in regions facing arsenic pollution, further complicates water safety and remediation efforts (Gupta et al. 2023). Every day, people from all walks of life in these parts of the world face this tasteless odourless threat by the simple act of using contaminated water. Understanding the scope of this issue is crucial, especially in economically impoverished areas, where integrated water resource management strategies can play a vital role in mitigating contamination risks (Yadav et al. 2024). These regions exhibit heightened arsenic contamination in groundwater, frequently coupled with a lack of alternative water sources or remediation methods, necessitating an urgent and cost-effective solution.

Although quite a few scalable solutions exist for the remediation of arsenic contamination of groundwater, such as ion exchange (Rathi et al. 2021), chemical oxidation (ALSamman et al. 2023), and precipitation (Ostermeyer et al. 2020). However, novel solutions employing photocatalytic processes (Raez et al. 2021) and other metallic nanoparticles (Pintor et al. 2018; Safi et al. 2019) have also been developed recently. However, these methods fall short of addressing the issue at hand, as they demand high startup costs, frequent maintenance, and a skilled workforce. Moreover, they scarcely address the problem of safe disposal of the contaminants removed from water and the long-term sustainability of the process. Among the techniques developed for arsenic removal from aqueous solutions, adsorption shows a clear advantage in terms of cost-effectiveness and reliability. Scientists are exploring affordable and sustainable materials ranging from microorganisms to industrial waste that show potential to act as adsorbents. Several of these options are already available on the market and some are yet to be optimized, with the scope of cost-effective, eco-friendly adsorbents for arsenic remediation lying just beyond the horizon (John et al. 2018; Rahidul Hassan 2023).

9.2 Adsorption for Arsenic Removal

Adsorption is a method of removal of particles from liquid or gas solutions by adherence to certain materials. These materials capable of attracting other compounds on their surface are called adsorbents. Adsorption is thus a surface phenomenon attributed to the action of Van der Waals forces and the electrostatic forces between the adsorbate and the adsorbent surface. Among all the methods used for arsenic removal, adsorption is the most discussed, with rapid advancements in removal efficiency and adsorbent synthesis taking place. This interest can be ascribed to the very nature of adsorption, which provides high removal efficiency with simple operation, low economic demand, and low impact on the environment. The flexibility of material choice and simple modification techniques attract more researchers into contributing to its development. Currently, adsorption without prior chemical addition is being reported to have a quantitative efficiency of around 95% for the remediation of both arsenite and arsenate forms. The parameters that show a notable influence on adsorption efficiency are temperature, pH, initial concentration, adsorbent dosage, and exposure time. Optimization of these parameters is conducted for each new adsorbent that is synthesized and metered accordingly. In general, arsenate adsorption is optimized at low pH levels and arsenite is adsorbed at pH ranges of 4–9. Removal efficiency increases with the increase in initial arsenic concentration in most of the adsorbents, especially observed in biobased adsorbents (John et al. 2018).

After the recognition of arsenic contamination as a global health threat, the adjoining objective of environmental scientists and researchers in related fields has been to find an effective, affordable, and reliable method for arsenic remediation from water. Affordability is of high significance because more than half of the regions that are critically affected by arsenic contamination are the ones from economically backward, resource-constrained regions. Therefore, mere removal efficiency is not sufficient for the technique to be feasible. Materials of natural origin with negligible or no processing have been studied by native researchers for immediate solutions. Even when not always producing significant results, such contributions have inspired research and development of the variety of low-cost adsorbents that are currently in the market. The unfortunate statistics concerning arsenic contamination levels worldwide indicate a correlation between higher risk and regions with lower economic capacity. More than half of the population currently showing explicit symptoms of arsenicosis comes from low-income backgrounds, increasing the severity of the issue. Limited economic resources lead to reduced access to proper healthcare care and even less access to effective arsenic removal techniques for water sources. In the absence of alternative water sources, the unprivileged are forced to consume arsenic contaminated water. Therefore, to protect millions of lives at risk of arsenic exposure and improve the quality of life for people across the world, reliable and affordable methods for arsenic decontamination are required. These methods must not only be cost effective but also be sustainable in the long term, considering the limited and ever-depleting resources. Thus, the requirement for the choice of remediation techniques that are environmentally benign and accessible to all communities lay the foundation for the immediate significance of developing low-cost adsorbents for the removal of arsenic (Podgorski and Berg 2020).

9.3 Types of Low-Cost Adsorbents

Adsorption is a simpler and more versatile method for arsenic remediation, and in order to further increase its feasibility, low-cost alternatives were developed. The reduction in cost indirectly caused a few diversions in the synthesis techniques of the adsorbents. The first diversion was the shift in interest into naturally derived or biobased raw materials to act as adsorbents and the second change includes the utilization of waste materials as potential adsorbents. Both of these diversions imply that a majority of the currently feasible low-cost adsorbents are not only economical but also cause less damage than its chemically processed alternatives. A wide variety of sources have been used for the synthesis of potential adsorbents for arsenic, including zeolites, activated alumina, chitosan, fly ash, rice husk, egg shell, activated carbon, and even chicken feathers. These varieties of materials can be broadly classified on the basis of their derived source as mineral-based, industrial waste-based, and bio-based. Each of these classifications and major examples from each are discussed below.

9.3.1 Mineral-Derived Adsorbents

9.3.1.1 Activated Alumina

Alumina or Aluminum oxide is a naturally occurring compound of aluminum denoted by the chemical formula (Al_2O_3). In its activated state, it acts as an effective adsorbent for the removal of arsenic from water, which is usually carried out by passing the contaminated water through packed reactors containing activated alumina. Activated alumina has shown above 90% removal of As(III) in the neutral pH range, and the adsorption is noted to be highly influenced by adsorbent dosage, contact time, and pH. The surface charge of the adsorbent and the form in which arsenic species are present also govern the arsenic adsorption onto activated alumina. It is reported that arsenite adsorption by activated alumina follows first-order kinetics and is highly dependent on pH and temperature. The adsorption of arsenic on activated alumina fits into Langmuir isotherm and is proved to be used as an effective adsorbent for arsenic removal. For comparison, the adsorption capacities for As(III) by unmodified alumina and activated alumina have been reported as 0.925 mg/g, and 3.48 mg/g respectively, with the latter showing a significant increase. It is also noted that activated alumina shows higher adsorption capacity with comparatively lower initial concentrations. The affordability of activated alumina due to its availability and cheap processing retains its position as a low-cost adsorbent for heavy metal removal (Rahidul Hassan 2023).

9.3.1.2 Zeolites

Zeolites refer to a group of crystalline microporous aluminosilicate materials that contain alkali or alkaline earth metals. Natural zeolites are often formed from sedimentary or igneous rocks, and synthetic zeolites are usually synthesized by the slow crystallization of silica alumina gel. Both synthetic zeolites and modified natural zeolites have been studied as adsorbents for the removal of arsenic. Synthetic zeolites have also shown a reliable arsenate removal capacity over a wide range of pH values. These can be especially useful because of their pH buffering capacity, which will reduce the cost and process time for effluent treatment by avoiding preconditioning. Modifications of the zeolite particles by coating with materials such as chitosan have also shown an increase in the adsorptive capacity. Thus, among the mineral-derived adsorbents suitable for the removal zeolite remains to be a cost-effective and efficient choice (Han et al. 2019).

9.3.1.3 Clay Minerals

Clay has been widely used as an adsorbent to remove various pollutants in different regions due to its accessibility and low cost. The development of effective adsorbents of clay constituents was hindered mainly by its particle size and behavior in aqueous mediums. Clay in its powdered form is often hard to remove after adsorption, and natural clays that appear larger in size break down or lose form while in contact with water for long periods. Therefore, its potential applications for the removal of heavy metals such as arsenic in dynamic flow systems were limited. However, cost-effective, simple modification methods, such as impregnation or iron coating, were found to significantly improve the removal efficiency and stability of clay-based adsorbents. Clays calcined in ferric and ferrous solutions can now be easily produced to remove substantial amounts of arsenate from water (Te et al. 2017). Laterite and kaolinite samples with negligible modifications are also utilized for arsenic adsorption. Arsenic adsorption is also aided by the relatively high aluminum and iron presence within the laterite. Laterites on treatment with acids and kaolinite modified with hexadecyl-trimethylammonium bromide (HDTMA-Br) have shown promising adsorption capacities. Laterite has shown higher efficiency in arsenic removal and exhibits maximum removal in comparison with kaolinite. Adsorption studies on various clay minerals have collectively found that overall arsenic adsorption is a combined effect of physical adsorption and chemisorption (Rehman et al. 2022).

9.3.2 Industrial and Agricultural Waste-Based

9.3.2.1 Metal and Blast Furnace Slags

Industrial waste from metal industries particularly from iron and steel manufacturing processes have been developed as a cost-friendly adsorbent for arsenic. The mechanism of arsenic removal is attributed to its chemical affinity for iron oxides in hydrolyzed forms. Metal loads, furnace linings, ash, and fluxing agents together form a mixed waste called metallurgical slag or blast furnace slag (BFS). These slag mixtures contain iron oxides as their main constituents. When hydrolysis occurs in contact with an aqueous solution containing arsenic, active adsorption sites form on the iron slag due to the formation of oxyhydroxides. Hydroxide ions and ions present in the arsenic oxides undergo a nucleophilic reaction on the surface of the adsorbent. These metal slags have produced effluents of up to 0.025 mg/L but are still to be improved in terms of adsorption capacity (Mercado-Borrayo et al. 2014). BFS is inexpensive, widely available, and readily contains high levels of iron which are superior for the removal of arsenic. It is also environmentally benign as the metals being tightly bound in structures that undergo negligible leaching, thus emerging as a promising source of an adsorbent.

9.3.2.2 Agricultural Waste

Readily available agricultural waste, such as rice husk, wheat straw, peanut shells, and tea waste, has been synthesized as adsorbents for the decontamination of As-containing water. The efficiency of agricultural waste is highly dependent on its type, initial concentration, and pH level. Higher initial concentrations and lower pH levels (5–6) favor higher arsenic removal by agricultural waste. Up to 92% removal efficiency was reported for 400 μg/L of initial arsenic concentration (Nagra et al. 2023). The superiority of these agricultural wastes being modified as low-cost adsorbents lies mainly in the their lower impact on the environment and waste utilization. In addition to rice husks, rice bran, and rice straw can also be synthesized for the adsorption of arsenic ions. In arsenic adsorption studies on rice husk-based adsorbents, an increase of 85% removal was observed under optimum conditions of pH 6. The lowest adsorption capacities were obtained with rice straw and rice bran ranging between 25.6 and 79.0 g/g and the highest adsorption capacity was reported with iron oxide amended with rice husk nanoparticles with a maximum value of 82 mg/g (Collivignarelli et al. 2022).

9.3.2.3 Biochar

Biochar refers to black carbon produced from plant residues, wood chips, manure, and other biomass sources. It is a more stabilized form of biomass carbon often prepared by pyrolysis of biomass (Sahoo et al. 2024). It has traditionally been used for carbon sequestration and soil nutrition and recently has been developed as a potential adsorbent for the removal of arsenic from water. Biochar offers a high surface area for adsorption and can be modified using nickel or manganese to improve its adsorption capacities. The special feature of biochar lies in its capacity to change surface charges with respect to pH levels. High removal rates have been achieved by impregnation of metals such as Fe, Zn, and magnetic modification (Siddiq et al. 2022). Biochar shows promise as a potential low-cost adsorbent for arsenic with its high porosity and excellent sorbent capacities. According to Collivignarelli et al. (2022), in its unmodified state, biochar derived from rice husk has shown an adsorption capacity greater than 19.3 mg/g. However, it is understood that there is inadequate development in its functionality to be employed in large-scale treatment facilities as of now.

9.3.3 Biosorbents

Biosorbents are adsorptive materials sourced from biological origins. Biosorbents in their raw form show less adsorptive capacities compared to their synthetic counterparts but are often cost-effective and environmentally benign. Biological materials that are rich in surface functional groups along with a highly porous structure possess high potential as active adsorbents with the right modifications. Several indigenous biosorbents, typically plant parts and agricultural waste, are of interest for research in regions with economic and technological constraints. Chitosan-based biosorbents have also been widely synthesized because of the inherent affinity biopolymers have been found to show a natural affinity toward transition metal irons. Similar to chitosan, melanin functionalized with copper and iron impregnation was found capable of removing more than 99% of As(III) and As(V) from water. An optimal pH range is between 4 and 6 for this rate of removal (Manirethan et al. 2020).

Bioremediation techniques that utilize organisms such as algae, bacteria, and algae have shown a capacity to remove the arsenic content from the aqueous solution as well. Fungal biomass and microalgae clusters act as adsorbent surfaces for arsenic ions causing effective removal. Macrophytes and several aquatic florae are also being studied for their ability to adsorb arsenic naturally. Due to the risks associated with the synthesizing of microorganisms as adsorbents, the development of biosorbents has shifted toward those derived from abundant, locally available natural materials. These easily accessible biosorbents offer a promising solution to arsenic remediation, making it more attainable for communities in developing regions of the world. A potato peel- and rice husk-derived adsorbent termed PPRH-ash was found to be applicable in natural groundwater with more than 90% adsorption efficiency. In addition to the fact that both raw materials are cheaply sourced, biosorption was observed to be selective for arsenic ions with minimal influence from other ions present in the sample (Bibi et al. 2017). Pilot-scale studies on arsenic adsorption in eggshell and java plum seeds have shown significant results and promising development as a biosorbent for removal of arsenic species. At neutral pH, the egg shell-based biosorbent is able to remove 87% of As(III) and 71% of As(V), and java plum seed removes 78% As(III) and 67% As(V) with very little pretreatment. Used plant parts such as watermelon peel, coconut shell, etc., have also been studied as potential adsorbents. In addition to designated plant parts being developed as arsenic adsorbents, specific mixtures of plant materials have also been reported to promise high removal efficiency. Indigenous biomass resources such as *Cucumis pubescens* (peel of fruit), *Moringa oleifera* (Lamarck), powdered leaves (Ashoka), lichen biomass, neem bark, coconut fibre, chicken feathers, animal fiber, sugarcane bagasse, sludge of various flowers like hibiscus and wide variety of wasted, abundant biomass materials are currently being developed into potent adsorbents for arsenic removal (Shakoor et al. 2019). Additionally relying on locally available biomass materials as sources of adsorbent material makes it a more feasible method for reducing arsenic contamination in areas that lack access to other expensive techniques. Much like chemically synthesized adsorbents, most of the currently established biosorbents display higher efficiency in removal of arsenite compared to that of arsenate.

9.4 Advancements in Adsorption Technologies

The recent surge in population has created an unprecedented demand for clean and safe drinking water, highlighting the need for cost-effective, rapidly deployable, and environmentally friendly water remediation methods. Advances in adsorption technology are enhancing water treatment capabilities and offering more effective and lasting arsenic remediation techniques. Material science and engineering breakthroughs have led to high-performance adsorbents with superior selectivity, capacity, and regeneration. Novel materials like functionalized biochars, hybrid composites, and enhanced nanomaterials present promising solutions to the limitations of conventional adsorbents. Some of these advances are detailed below.

9.4.1 Hybrid Adsorbents

Hybrid adsorbents such as biochar- and chitosan-based materials combine multiple materials to enhance arsenic removal efficiency, offering improved selectivity, capacity, and regeneration capabilities, making them a promising solution for effective water treatment.

Biochar-based hybrid adsorbents in addition to being a cost-effective and sustainable alternative to conventional arsenic adsorbents have the added benefit of being compatible and in most cases beneficial in use cases involving soil remediation due to its nature as a pH buffer and nutrient moderator (Hossain et al. 2020). An example of such a hybrid biochar-based adsorbent was developed and studied by Lin et al. (2019). Their study focused on the development and efficacy of biochar impregnated with Fe–Mn oxides that utilized biochar derived from powdered corn stem biomass. Base biomass was subject to slow pyrolysis in an N_2 environment, after which it was treated with $FeSO_4$ and $KMnO_4$ solutions according to weight ratio. Batch adsorption studies determined that the ideal weight ratio of biochar:Fe:Mn is 18:3:1. The BET surface area of the adsorbent in question was determined to be 7.53 m^2/g with an astonishing oxidation reduction potential of 317.1 mV. The adsorption process was found to conform best to the pseudo-second order kinetic model. The Freundlich isotherm model was reported to be the best fit, revealing a high bias towards the chemisorption process with an apparent adsorption capacity of 8.80 mg/g. Subsequent batch adsorption studies helped determine that having an initial concentration of As(III) of 30 mg/L at 3–7 pH, equilibrium was achieved in under 120 min with an over 92% reduction in As(III) content. Further, FTIR and XPS studies helped develop the mechanism of As(III) adsorption which was determined to occur primarily by 3 means namely, the oxidation of As(III) to As(V) by both Fe but primarily by Mn, the chemisorption of As(V) onto Fe and the minimal but direct adsorption of As(III) onto Fe. Similar studies conducted by Fan et al. (2020) using zero valent Fe-impregnated biochar and Wang et al. (2022) using sulfide/ferro-ferric oxide coated biochar to remediate arsenic-contaminated soils provided promising results.

Another class of hybrid adsorbents leverages the stability and robust framework of metallic oxide minerals, commonly combined with the biocompatibility and biodegradability of chitosan. This synergy enhances the efficiency of adsorption, making these hybrid materials highly effective for environmental remediation applications. A recent and novel example of this is explained in the work of Han et al. (2019) who developed and studied the efficacy of using chitosan-coated Na–X zeolite (C-NX) for the adsorptive removal of As(V). The adsorbent was prepared by first homogenizing and calcining fly ash with NaOH, which formed the matrix for the adsorbent, after which it was mixed into an acetic acid chitosan solution, which was then filtered and the adsorbent washed and dried. XRD and FTIR analysis were performed to determine the chemical composition and crystal structure of the resulting adsorbent, which confirmed the successful coating of chitosan on the Na–X zeolite surface. Batch adsorption studies carried out with an initial concentration of 5 mg/L at 2.1 pH showed a nearly complete removal of As(V) from the solution with an equilibrium time of 2 h. With a notable reported capacity of 63.23 mg/g, the sorption was shown to occur in line with the Langmuir adsorption isotherm, suggesting the dominance of monolayer adsorption, and to correspond to the pseudo-second-order kinetic model. XPS and FTIR analysis of the samples before and after adsorption helped determine the mechanism of adsorption. It was determined that protonated amino surface groups (–NH_3^+) of chitosan and the Al–OH played the greatest role in the adsorption by surface complex formation due to the large positive charge they facilitate. It should be noted that this adsorbent functions in very low pH ranges (2.1–4.6 pH), making it a perfect fit for remediating mining wastewater, which frequently requires adjustment of the pH prior to remediation.

9.5 Future Directions and Research Needs

Low-cost adsorbents have gained great popularity in wastewater treatment techniques. However, the prevailing limitations consisting of poor surface area and stability, lower mechanical characteristics, and difficulty in reusability have hindered the major functionality of these conventional adsorbents.

9.5.1 Pyrolysis

Pyrolysis is a process that, in the absence of oxygen, thermally decomposes biomass into its byproducts. This method is particularly significant for producing low-cost adsorbents from agricultural wastes and other organic wastes for the removal of arsenic. The novel methods discovered have been introduced to provide a more efficient, cost-effective, faster, and environmentally friendly approach. One such method is the pyrolysis self-activation of bio-waste into activated carbon. Gao et al. (2023) presented a technique for generating activated carbon by self-activated pyrolysis of bamboo shoot shells. Figure 9.1 shows the pictorial representation of the system. Self-activated pyrolysis conducts its process without utilizing any external activator or auxiliary gases, additionally, it makes use of the gaseous by-products of the process, namely the CO_2 and H_2O that are produced. These reagents also have a notably high heating value of 11.7 MJ/N m^3. Thus, formed activated carbon exhibited excellent arsenic adsorption capability with a maximum capacity of 10.9 mg/g for As(III). The process was carried out at an activation temperature of 10,000 C with 4 h of residence time. The study suggested that future research in reducing the residence period of the process as well as additional utilization of combustible gas, could overall improve the effectiveness and efficiency of the process.

Copyrolysis, a process in which more than one material is used as feedstock compared to conventional pyrolysis (Prakash et al. 2018). This technique is gaining popularity in the field of modification of low-cost adsorbents in arsenic removal. One such study is conducted by Wang et al. (2022), where the work used incinerated sewage sludge ash (ISSA) from iron, which contains solid waste derived after the combustion of dewatered sludge. This sludge was then co-pyrolyzed with peanut shell in order to produce an iron-based biochar that exhibited arsenic adsorption properties, 17.44 and 57.6 mg/g for As(III) and As(V), respectively. However, the

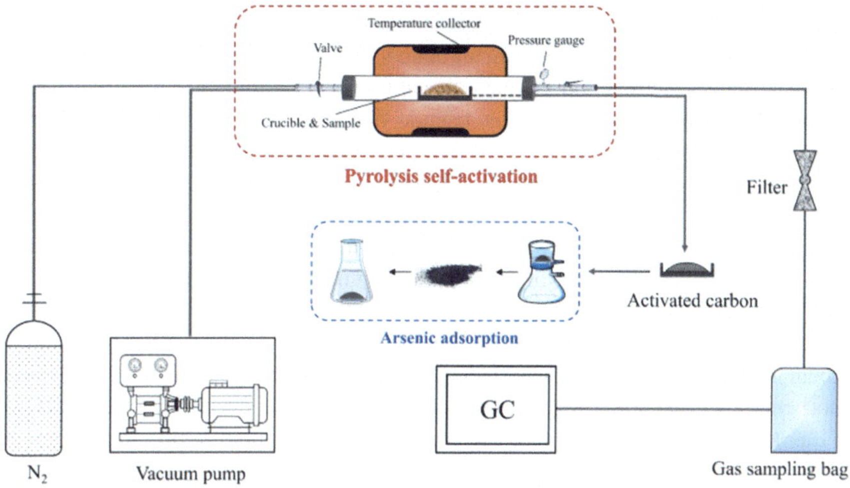

Fig. 9.1 Schematic representation of the self-activation pyrolysis of bamboo shoot shells into activated carbon for arsenic removal. Reprinted from Gao et al. (2023), with permission from Bioresource Technology. Copyright 2023, Elsevier

process displayed a drawback of high leaching of reactive silica from the biochar, which resulted in hindering active adsorption sites as well as corrosive product formation (Chen and Poon 2017). Even though the process was found to reduce production costs, further research should be conducted to tackle this problem. Microwave pyrolysis, a thermochemical process that necessitates microwave heating to obtain extensive decomposition, is another procedure that is gaining popularity in this field. The novel bagasse biochar/Ferrites nanocomposite produced by this method demonstrated not only a high performance adsorbent, but also a low cost methodology that widely used readily available renewable biomaterials (Bai et al. 2022). Plastic pyrolysis, the heating of plastic at extremely high temperatures with a lack of oxygen, has recently been used as an ecosynthesis method that employs waste plastic, revealing excellent surface chemistry along with the evolved morphology of activated char adsorbents (Mukherjee et al. 2024).

Modified versions of pyrolysis have shown a varied range of adsorbents. However, they have certain drawbacks. The scope of future studies is found to tackle these limitations to obtain the highest efficiency of these adsorbents, consequently making these processes more cost-effective.

9.5.2 *Hybridisation*

Hybridization process can be accomplished through the incorporation of multiple low-cost adsorbents in order to improve adsorption efficiency (Eniola et al. 2020).

Basu et al. (2021) developed a combination of microspheres of GO–MnO_2–Goethite core–shell impregnated onto calcium alginate to form hydrogel beads (Fig. 9.2). The work also details the reaction kinetics, thermodynamic parameters, and pH studies. Here, the beads were found to follow pseudo-first-order kinetics conforming to the isotherm model of the Langmiur isotherm and rather found to be endothermic and sponataneous in nature. This is attributed to the improvement in the rate of interparticle diffusion of arsenic groups into the modified beads adsorbent. Furthermore, it showed maximum uptake at a pH of 4–6. The evaluation indicated that 27.53 mg/g as the maximum sorption capacity for As(III) and the maximum adsorption capacity obtained for As(V) was 34.17 mg/g.

Similarly, Mudzielwana et al. (2020) synthesised a inorgano–organo hybrid clay sorbent through an intercalation surfactant cation in interlayers of kaolin clay. These adsorbents also displayed regenerative properties up to 7 cycles by using 0.01 M HCl as the regenerant. They were able to achieve maximum sorption capacity at room temperature with 7.99 mg/g and 7.32 mg/g for As(III) and As(V) respectively. Das et al. (2020) discovered an ultra-high adsorbent through the deposition of Fe nanoparticles onto GO, namely GFeN, had exceptionally high adsorption power by

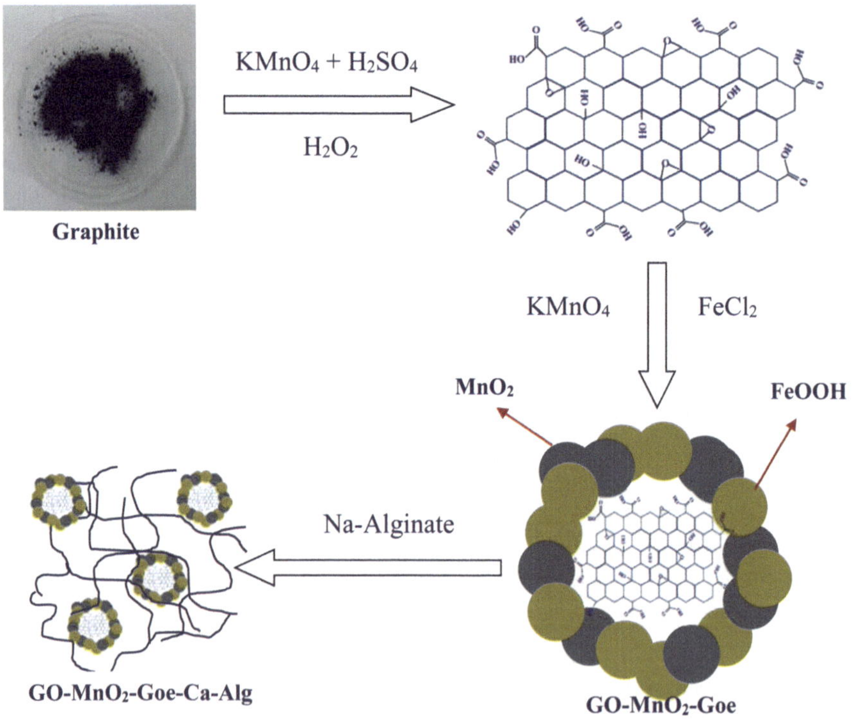

Fig. 9.2 Schematic representation of synthesis of of GO–MnO_2–goethite–Ca–alginate beads. Reprinted from Basu et al. (2021), with permission from *Journal of Water Process Engineering*. Copyright 2021, Elsevier

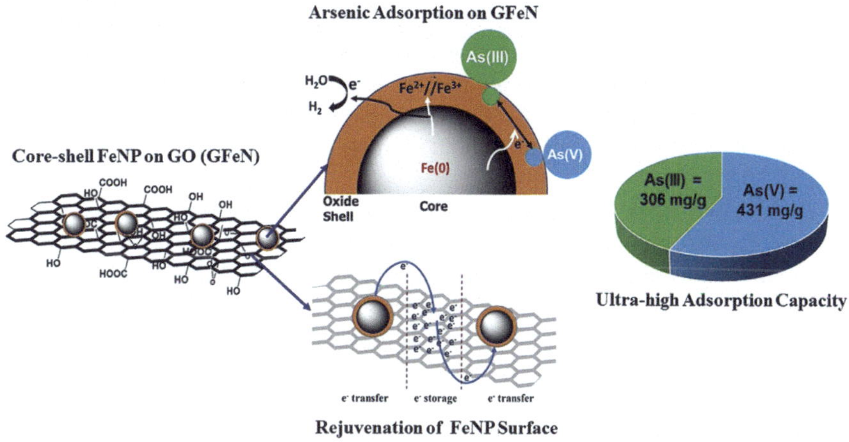

Fig. 9.3 Graphical representation of core–shell FeNP on GO for arsenic adsorption. Reprinted from Das et al. (2020), with permission from *Chemosphere*. Copyright 2020, Elsevier

improving Fe efficiency and arsenic mobility (Fig. 9.3). Electrostatic interaction and surface complexation interposed with this significantly high adsorption capacity. The mechanism of GO sheets modeled to be a reservoir for electron release as a result of corrosion of the FeNP surface and the electrons being transferred to rejuvenate the oxidized platform additionally helped in the removal of arsenic. The adsorption capacities for As(III) and As(IV) were obtained as 306 mg/g and 431 mg/g, respectively.

Higher removal rates of the heavy metal arsenic were observed after hybridization. Apart from its exceptional characteristics, there are inadequate data available on the eradication of microbes by these materials, which also have significant importance in water treatment (Baig et al. 2021). Further research should also focus on extending its applicability in the removal of microbial, organic, and inorganic residual pollutants.

9.5.3 Aerogel

Aerogels and their derivatives are emerging adsorbents in the field of heavy metal remediation. Aerogels are porous solid networks with a large majority of the space within the material occupied by airpockets. Remarkable characteristics including exceptionally low density, immense active surface area, and lower thermal conductivity make it to be suitable for fabrication on different materials (Ihsanullah et al. 2022). The cost of nanosynthesis and the complicated methods employed in the fabrication of aerogels are currently being compensated for by using renewable and cheaper materials for fabrication. One of the latest studies that developed a hierarchically monolithic porous Iron Metal–Organic Gel/Bacterial Cellulose Aerogel (Fe–MOG/BC) was fabricated with abundant active site, mechanically stability, and ultralight nature. It showed a maximum adsorption capacity of 495 mg/g for arsenate. In addition to its excellent adsorbing capacity, its ease of generation, low cost, high selectivity, scalability, and reusability favours it a good low cost adsorbent. Furthermore, the aerogel also enables effective decontamination of As(V) up to 5 ppm and was also potentially found to extend both in the point of use and the point of entry (Li et al. 2021) (Fig. 9.4).

Comparably, Yu et al. (2019) acquired Fe_2O_3 nanocubes that are impregnated onto a porous graphene aerogel that was found to have a maximum adsorption capacity of 172.27 and 217.34 mg/g for As(III) and As(V), respectively. Reduction of a highly concentrated solution of As(III) can be obtained within 60 min and reduction of As(V) with only 5 min, which displays an ultra-fast removal capability. Furthermore, even after 8 cycles, it was observed to have a very slight decline in terms of removal efficiency.

Thus, the aerogel is able to demonstrate a future of low-cost adsorbents, especially due to its ultra-fast and high adsorption power. The research scopes are based on experimentation with various naturally occurring materials and wastes along with

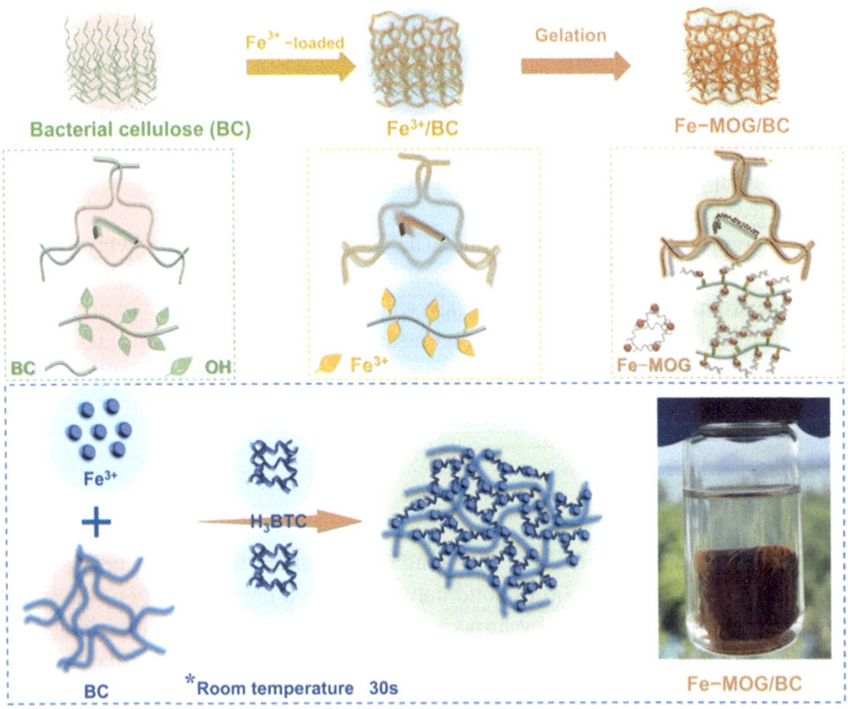

Fig. 9.4 Schematic diagram of fabrication of Fe–MOG/BC aerogel. Reprinted from Li et al. (2021), with permission from *ACS Applied Materials and Interfaces*. Copyright 2021, American Chemical Society

finding cost-effective fabrication methods, which can potentially create a new era of adsorbents.

Implementation of supportive policies and regulatory frameworks is crucial to advance the usage of these low-cost adsorbents. Regular monitoring through Life Cycle Assessment (LCA) and Health Risk Assessment must be performed to keep a check on the impact of these adsorbents. Encouraging research in the direction of regeneration and reusability is found to be equally important in developing a noval adsorbent. This could also contribute to a reduction in production costs. Hence, various factors affect the inclusive cost of an adsorbent in addition to the raw material used. Therefore, thorough research and implementation of policies and recommendations are vital in establishing efficient mitigation strategies for combating arsenic contamination.

9.6 Challenges and Limitations

Efforts to develop low-cost adsorbents for arsenic removal show promise, but face significant limitations. Each synthesis method varies, leading to inconsistent efficiency. For example, a study by Nagra et al. (2023) found that tea waste, wheat straw, and peanut shells had varying arsenic removal efficiencies influenced by initial concentration and pH levels. Most adsorbents perform best at neutral pH but struggle with fluctuations, and efficiency drops with lower arsenic concentrations, limiting their use for moderate or low-contaminated samples. Additional challenges include poor selectivity, interference from other ions, and issues with regeneration and reusability. Adsorbents often need replacement after four to five regenerations, making them less efficient than other methods. The absence of self-monitoring in these systems raises concerns about treatment quality, especially in regions with limited access to skilled labor. Biological adsorbents have lower reusability and pose environmental risks due to arsenic-rich waste. Laboratory results often fail to match real-world conditions. While modifications such as the incorporation of nanoparticles show potential, cost and complexity remain barriers to large-scale application. Although it is cheaper and more environmentally friendly, commercialization faces challenges due to production costs and infrastructure needs. The disposal of arsenic-rich spent adsorbents poses environmental risks, and reliance on region-specific materials complicates replication elsewhere, potentially leading to bioresource overexploitation.

9.7 Conclusions

Arsenic contamination has been regarded as one of the main concerns due to its high toxicity even at lower concentrations, to environment and health. When compared to conventional removal techniques, adsorption is an extensively used technique primarily because of its easy usage, efficiency, and cost-effectiveness. Moreover, the regeneration of these adsorbents also plays an important role in determining the performance of the process. To obtain an optimised functionality of the process, low-cost and highly effective adsorbents are an advancement still ongoing. Recent research has highlighted various promising adsorbents including bioadsorbents, agricultural and industrial waste, metal-doped oxides and composites, and hybrid adsorbents, which offer unique benefits that contribute to viable solutions for arsenic remediation. The development and modification of these adsorbents are promising in addressing the contamination issues. Emerging low-cost adsorbents consisting of hybrid adsorbents and aerogels, as well modified techniques of existing methodology, have been found to be reassuring. However, these existing materials or fabrication techniques are often observed to have certain drawbacks. Therefore, extensive research and innovation in this areana are essential to optimize the performance, usability, scalability, and environementally friendly method, thus ensuring a reduction in arsenic contamination.

Acknowledgements The authors thank our prestigious institute and other people who are involved in this study for their direct and indirect contributions.

References

Adeloju, S. B., Khan, S., & Patti, A. F. (2021). Arsenic Contamination of Groundwater and Its Implications for Drinking Water Quality and Human Health in Under-Developed Countries and Remote Communities—A Review. Applied Sciences 2021, Vol. 11, Page 1926, 11(4), 1926. https://doi.org/10.3390/APP11041926

ALSamman, M. T., Sotelo, S., Sánchez, J., & Rivas, B. L. (2023). Arsenic oxidation and its subsequent removal from water: An overview. Separation and Purification Technology, 309, 123055. https://doi.org/10.1016/J.SEPPUR.2022.123055

Bai, L., Huang, A., Feng, J., Su, X., Mo, W., Ma, S., & Lin, H. (2022). One-step microwave pyrolysis synthesis of bagasse biochar/ferrites nanocomposite and synergistic effect on As(V) adsorption in water. Materials Chemistry and Physics, 283, 126035. https://doi.org/10.1016/j.matchemphys.2022.126035

Baig, U., Faizan, M., & Sajid, M. (2021). Effective removal of hazardous pollutants from water and deactivation of water-borne pathogens using multifunctional synthetic adsorbent materials: A review. Journal of Cleaner Production, 302, 126735. https://doi.org/10.1016/j.jclepro.2021.126735

Basu, H., Singh, S., Venkatesh, M., Pimple, M. V., & Singhal, R. K. (2021). Graphene oxide-MnO_2-goethite microsphere impregnated alginate: A novel hybrid nanosorbent for As (III) and As (V) removal from groundwater. Journal of Water Process Engineering, 42, 102129. https://doi.org/10.1016/j.jwpe.2021.102129

Bibi, S., Farooqi, A., Yasmin, A., Kamran, M. A., & Niazi, N. K. (2017). Arsenic and fluoride removal by potato peel and rice husk (PPRH) ash in aqueous environments. International Journal of Phytoremediation, 19(11), 1029–1036.

Chen, Z., & Poon, C. S. (2017). Comparing the use of sewage sludge ash and glass powder in cement mortars. Environmental Technology, 38(11), 1390–1398. https://doi.org/10.1080/09593330.2016.1230652

Chutia, P., Kato, S., Kojima, T., & Satokawa, S. (2009). Arsenic adsorption from aqueous solution on synthetic zeolites. Journal of Hazardous Materials, 162(1), 440–447. https://doi.org/10.1016/J.JHAZMAT.2008.05.061

Collivignarelli, M. C., Sorlini, S., Milanese, C., Illankoon, W. A. M. A. N., Caccamo, F. M., & Calatroni, S. (2022). Rice Industry By-Products as Adsorbent Materials for Removing Fluoride and Arsenic from Drinking Water—A Review. In Applied Sciences (Switzerland) (Vol. 12, Issue 6). https://doi.org/10.3390/app12063166

Das, T. K., Sakthivel, T. S., Jeyaranjan, A., Seal, S., & Bezbaruah, A. N. (2020). Ultra-high arsenic adsorption by graphene oxide iron nanohybrid: Removal mechanisms and potential applications. Chemosphere, 253, 126702. https://doi.org/10.1016/j.chemosphere.2020.126702

Eniola, Jamiu., Kumar, R., A. Barakat, M., O. Eniola, Jamiu., Kumar, R., & A. Barakat, M. (2020). Nanomaterials for the Adsorptive Removal of Antibiotics from Aqueous Solution: A Minireview. General Chemistry, 6(2), 190018–190018. https://doi.org/10.21127/yaoyigc20190018

Fan, J., Chen, X., Xu, Z., Xu, X., Zhao, L., Qiu, H., & Cao, X. (2020). One-pot synthesis of nZVI-embedded biochar for remediation of two mining arsenic-contaminated soils: Arsenic immobilization associated with iron transformation. Journal of Hazardous Materials, 398. https://doi.org/10.1016/J.JHAZMAT.2020.122901

Gao, Q., Feng, Z., He, Y., Hou, Y., Ren, H., Su, M., Ni, L., & Liu, Z. (2023). Pyrolysis self-activation: An environmentally friendly method to transform biowaste into activated carbon for arsenic removal. Bioresource Technology, 368, 128353. https://doi.org/10.1016/j.biortech.2022.128353

Gupta, S., Mishra, N., Kumar, A., Yadav, A.K. (2023). Fluoride Pollution in Subsurface Water: Challenges and Opportunities. In: Yadav, A.K., Shirin, S., Singh, V.P. (eds) Advanced Treatment Technologies for Fluoride Removal in Water. Water Science and Technology Library, vol 125. Springer, Cham. https://doi.org/10.1007/978-3-031-38845-3_2

Han, C., Yang, T., Liu, H., Yang, L., & Luo, Y. (2019). Characterizations and mechanisms for synthesis of chitosan-coated Na–X zeolite from fly ash and As(V) adsorption study. Environmental Science and Pollution Research, 26(10), 10106–10116. https://doi.org/10.1007/S11356-019-04466-X

Hossain, M. Z., Bahar, M. M., Sarkar, B., Donne, S. W., Ok, Y. S., Palansooriya, K. N., Kirkham, M. B., Chowdhury, S., & Bolan, N. (2020). Biochar and its importance on nutrient dynamics in soil and plant. Biochar 2020 2:4, 2(4), 379–420. https://doi.org/10.1007/S42773-020-00065-Z

Ihsanullah, I., Sajid, M., Khan, S., & Bilal, M. (2022). Aerogel-based adsorbents as emerging materials for the removal of heavy metals from water: Progress, challenges, and prospects. In Separation and Purification Technology (Vol. 291, p. 120923). https://doi.org/10.1016/j.seppur.2022.120923

John, Y., David, V. E., & Mmereki, D. (2018). A Comparative Study on Removal of Hazardous Anions from Water by Adsorption: A Review. In International Journal of Chemical Engineering (Vol. 2018). Hindawi Limited. https://doi.org/10.1155/2018/3975948

Kuivenhoven, M., & Mason, K. (2023). Arsenic Toxicity. StatPearls. https://www.ncbi.nlm.nih.gov/books/NBK541125/

Kumar, A., Chava, R., Gupta, S., Shirin, S., Jamal, A., Yadav, A.K. (2023). Environmental and Health Effects of Fluoride Contamination and Treatment of Wastewater Using Various Technologies. In: Yadav, A.K., Shirin, S., Singh, V.P. (eds) Advanced Treatment Technologies for Fluoride Removal in Water. Water Science and Technology Library, vol 125 . Springer, Cham. https://doi.org/10.1007/978-3-031-38845-3_18

Kumarathilaka, P., Seneweera, S., Meharg, A., & Bundschuh, J. (2018). Arsenic speciation dynamics in paddy rice soil-water environment: sources, physico-chemical, and biological factors - A review. Water Research, 140, 403–414. https://doi.org/10.1016/J.WATRES.2018.04.034

Li, H., Ye, M., Zhang, X., Zhang, H., Wang, G., & Zhang, Y. (2021). Hierarchical Porous Iron Metal-Organic Gel/Bacterial Cellulose Aerogel: Ultrafast, Scalable, Room-Temperature Aqueous Synthesis, and Efficient Arsenate Removal. ACS Applied Materials and Interfaces, 13(40), 47684–47695. https://doi.org/10.1021/acsami.1c14938

Lin, L., Song, Z., Huang, Y., Khan, Z. H., & Qiu, W. (2019). Removal and Oxidation of Arsenic from Aqueous Solution by Biochar Impregnated with Fe-Mn Oxides. Water, Air, and Soil Pollution, 230(5). https://doi.org/10.1007/s11270-019-4146-5

Manirethan, V., Raval, K., & Balakrishnan, R. M. (2020). Adsorptive removal of trivalent and pentavalent arsenic from aqueous solutions using iron and copper impregnated melanin extracted from the marine bacterium Pseudomonas stutzeri. Environmental Pollution, 257. https://doi.org/10.1016/j.envpol.2019.113576

Mercado-Borrayo, B. M., Schouwenaars, R., Litter, M. I., Montoya-Bautista, C. V, & Ramírez-Zamora, R. M. (2014). Metallurgical slag as an efficient and economical adsorbent of arsenic. In Water reclamation and sustainability (pp. 95–114). Elsevier.

Mudzielwana, R., Gitari, M. W., & Ndungu, P. (2020). Enhanced As(III) and As(V) Adsorption From Aqueous Solution by a Clay Based Hybrid Sorbent. Frontiers in Chemistry, 7. https://doi.org/10.3389/fchem.2019.00913

Mukherjee, A., Ruj, B., Sadhukhan, A. K., Gupta, P., Chakrabortty, S., Chatterjee, R., & Bishayee, B. (2024). Eco-synthesis, characterization, and application of waste plastics pyrolysis char in arsenic removal from contaminated water: An integrated circular framework with parametric response surface methodology optimization-cum-artificial neural network model. Journal of Environmental Chemical Engineering, 12(1), 111824. https://doi.org/10.1016/j.jece.2023.111824

Nagra, M. A., Natasha, N., Bibi, I., Tariq, T. Z., Naz, R., Ansar, S., Shahid, M., Murtaza, B., Imran, M., & Khalid, M. S. (2023). Biowaste-based sorbents for arsenic removal from aqueous medium and risk assessment. Environmental Geochemistry and Health, 45(12), 9017–9028.

Ostermeyer, P., Bonin, L., Folens, K., Verbruggen, F., García-Timermans, C., Verbeken, K., Rabaey, K., & Hennebel, T. (2020). Effect of speciation and composition on the kinetics and precipitation of arsenic sulfide from industrial metallurgical wastewater. https://doi.org/10.1016/j.jhazmat.2020.124418

Pintor, A. M. A., Vieira, B. R. C., Santos, S. C. R., Boaventura, R. A. R., & Botelho, C. M. S. (2018). Arsenate and arsenite adsorption onto iron-coated cork granulates. Science of the Total Environment, 642, 1075–1089. https://doi.org/10.1016/J.SCITOTENV.2018.06.170

Podgorski, J., & Berg, M. (2020). Global threat of arsenic in groundwater. Science, 368(6493), 845–850.

Prakash, R., Siddharth, R., & Gunasekar, N. (2018). Cracking of toxic waste. In Energy from Toxic Organic Waste for Heat and Power Generation (pp. 139–149). Elsevier. https://doi.org/10.1016/B978-0-08-102528-4.00010-9

Raez, J. M., Arencibia, A., Segura, Y., Arsuaga, J. M., & López-Muñoz, M. J. (2021). Combination of immobilized TiO_2 and zero valent iron for efficient arsenic removal in aqueous solutions. Separation and Purification Technology, 258. https://doi.org/10.1016/j.seppur.2020.118016

Rahidul Hassan, H. (2023). A review on different arsenic removal techniques used for decontamination of drinking water. In Environmental Pollutants and Bioavailability (Vol. 35, Issue 1). https://doi.org/10.1080/26395940.2023.2165964

Rathi, B. S., Kumar, P. S., Ponprasath, R., Rohan, K., & Jahnavi, N. (2021). An effective separation of toxic arsenic from aquatic environment using electrochemical ion exchange process. Journal of Hazardous Materials, 412, 125240. https://doi.org/10.1016/J.JHAZMAT.2021.125240

Rehman, U. ur, Khan, S., & Muhammad, S. (2020). Ingestion of Arsenic-Contaminated Drinking Water Leads to Health Risk and Traces in Human Biomarkers (Hair, Nails, Blood, and Urine), Pakistan. Exposure and Health, 12(2), 243–254. https://doi.org/10.1007/S12403-019-00308-W

Rehman, A., Rukh, S., Ayoubi, S. Al, Khattak, S. A., Mehmood, A., Ali, L., Khan, A., Malik, K. M., Qayyum, A., & Salam, H. (2022). Natural clay minerals as potential arsenic sorbents from contaminated groundwater: equilibrium and kinetic studies. International Journal of Environmental Research and Public Health, 19(23), 16292.

Safi, S. R., Gotoh, T., Iizawa, T., & Nakai, S. (2019). Development and regeneration of composite of cationic gel and iron hydroxide for adsorbing arsenic from ground water. Chemosphere, 217, 808–815. https://doi.org/10.1016/J.CHEMOSPHERE.2018.11.050

Sahoo, K., Kumar, A., Sonwani, R. K., Sonawane, S. H., & Chakraborty, J. P. (2024). A new insight into fixed-bed catalytic pyrolysis of sugarcane bagasse: optimization of process parameters and analysis of products. *Biomass Conversion and Biorefinery*, 1–22. https://doi.org/10.1007/S13399-024-05978-4/METRICS

Shakoor, M. B., Niazi, N. K., Bibi, I., Shahid, M., Saqib, Z. A., Nawaz, M. F., Shaheen, S. M., Wang, H., Tsang, D. C. W., & Bundschuh, J. (2019). Exploring the arsenic removal potential of various biosorbents from water. Environment International, 123, 567–579.

Siddiq, O. M., Tawabini, B. S., Soupios, P., & Ntarlagiannis, D. (2022). Removal of arsenic from contaminated groundwater using biochar: a technical review. International Journal of Environmental Science and Technology, 1 14.

Te, B., Wichitsathian, B., & Yossapol, C. (2017). Adsorptive behavior of low-cost modified natural clay adsorbents for arsenate removal from water. International Journal of GEOMATE, 12(33). https://doi.org/10.21660/2017.33.2531

Wang, G., Peng, C., Tariq, M., Lin, S., Wan, J., Liang, W., Zhang, W., & Zhang, L. (2022). Mechanistic insight and bifunctional study of a sulfide Fe_3O_4 coated biochar composite for efficient As(III) and Pb(II) immobilization in soils. Environmental Pollution, 293, 118587. https://doi.org/10.1016/J.ENVPOL.2021.118587

Yadav, A. K., Yadav, K., & Singh, V. P. (Eds.). (2024). Integrated Management of Water Resources in India: A Computational Approach. 129. https://doi.org/10.1007/978-3-031-62079-9

Yu, X., Wei, Y., Liu, C., Ma, J., Liu, H., Wei, S., Deng, W., Xiang, J., & Luo, S. (2019). Ultrafast and deep removal of arsenic in high-concentration wastewater: A superior bulk adsorbent of porous Fe_2O_3 nanocubes-impregnated graphene aerogel. Chemosphere, 222, 258–266. https://doi.org/10.1016/j.chemosphere.2019.01.130

Chapter 10
Low-Cost Arsenic Removal Using Different Biological Routes

Arham Tater, Siyaram Meena, Anamika Barman, Jyotirmay Roy, Sougata Roy, and Suman Dutta

Abstract Arsenic (As) contamination poses significant environmental and public health challenges due to its widespread occurrence and toxicity. Various sources and the extent of As contamination in soils exist that emphasize the urgent need for effective remediation strategies. Traditional and innovative technologies for As removal are explored, focusing on clay modification techniques, microbial remediation, and advanced oxidation reduction processes. The adsorptive ability of As is increased by clay modification, providing an affordable and long-lasting solution. Through procedures that include volatilization, methylation, bioaccumulation, and biosorption, some bacterial and fungal strains that are capable of transformation and detoxification are used in microbial remediation. Promising results have been observed in the reduction of As levels in polluted soils by merging biotechnological technologies with traditional procedures. The study also emphasizes recent developments in the mechanisms by which plants and microbes absorb and convert As, providing information on the development of more effective and environmentally benign remediation techniques. Understanding the interplay between these processes is crucial for optimizing remediation efforts and mitigating the undesirable influence of As contamination on ecosystem and human well-being.

Keywords Contamination · Soil remediation · Clay modification · Microbial remediation · Genetic modification

A. Tater · S. Meena · J. Roy
Division of Soil Science and Agricultural Chemistry, ICAR-Indian Agricultural Research Institute, New Delhi, India

A. Barman
Division of Agronomy, ICAR-Indian Agricultural Research Institute, New Delhi, India

S. Roy
International Potato Centre, New Delhi, India

S. Dutta (✉)
Ramakrishna Mission Vivekananda Educational and Research Institute, Narendrapur, India
e-mail: duttasumansiliguri@gmail.com

A. K. Yadav et al. (eds.), *Arsenic Removal Technologies*,
https://doi.org/10.1007/978-3-031-91171-2_10

10.1 Introduction

Arsenic (As) has emerged as a prominent threatening contaminant in recent decades. It is metalloid in its chemical nature, but it has been appropriately characterized as a heavy metal, due to its extreme harmfulness and lack of natural attenuation (Ahmed et al. 2022; Yadav et al. 2014). The menace of As is prevalent both in soil and groundwater, while the hazard risk is more associated with the latter case due to its higher availability. Reports indicate that almost 22 million people are directly or indirectly at risk around the world due to its hazardous adverse health effects (Podgorski and Berg 2020). It has the potential to adversely affect multiple organs and organ systems of humans such as the skin, kidney, circulatory system, nervous system, and reproductive system, and is also a well-established Group I carcinogen (Sarkar et al. 2022). Therefore, it is classified as a dangerous toxic element and it is extremely important to monitor and control the amount of As in food or water (Liao et al. 2022). For this purpose, the World Health Organization (WHO) has prescribed the maximum permissible limit of As in drinking water as $10\,\mu g\,L^{-1}$ (WHO 2022). In view of the fact that arsenic pollution has expanded widely, appropriate and effective mitigation strategies are therefore needed to reduce the negative impacts on human health and ecosystem.

This chapter aims to explore low-cost biological approaches for As removal, emphasizing their effectiveness, sustainability, and environmental impact. It provides a comprehensive analysis of As contamination sources, its health and environmental implications, and the need for efficient remediation strategies. The study examines various biological routes, including clay modification techniques, microbial remediation, and advanced oxidation–reduction processes, highlighting their potential to enhance As adsorption, transformation, and detoxification. Additionally, it inspects the role of microbial strains in As volatilization, methylation, bioaccumulation, and biosorption, as well as the collaborative effects of biotechnological innovations with traditional remediation methods. By understanding the mechanisms underlying plant and microbial As uptake and conversion, the chapter aims to contribute to the development of cost-effective, sustainable, and ecologically sound solutions for mitigating As contamination in soil and water systems.

10.2 Sources of As in the Soil

The sources of As in soil can be categorized as natural (due to geogenic reasons) or anthropogenic (for human-induced causes) (Wang et al. 2021). Natural sources include volcanic eruptions, thermal springs, and weathering of natural rocks and minerals enriched with As. Since minerals are found in varying proportions in the environment, the extent of As menace is dictated by the dominance of some minerals that specifically contain As in higher proportion. For example, the most abundant mineral in nature is arsenopyrite (FeAsS), while other minerals such as realgar (AsS) and orpiment (As_2S_3) are also recognized as important contributors to natural arsenic in the environment (Liao et al. 2022). On the other hand, anthropogenic sources are regional or site-specific human interventions that lead to build-up of As in the environment. These activities include mining and smelting operations, waste disposal, coal burning, sewage irrigation, and the application of agricultural inputs such as pesticides and fertilizers in the soil (Biswas et al. 2019; Shirin and Yadav 2021).

10.3 Extent of As Contamination

The average concentration of As in the baseline soil ranges from 5 to 10 mg kg^{-1} in the soil. Although the As concentration in Asian regions can escalate up to >50 mg kg^{-1} soil, the average concentration in Europe is just around 7 mg kg^{-1} soil. One of the main causes for higher As content in Asian regions (especially in Southeast Asia) can be attributed to irrigation of crops with As-burdened groundwater (Podgorski and Berg 2020); more specifically, the part of the Indian subcontinent serving as the catchment basin of the Ganga–Brahmaputra–Meghna river, which constitutes the Bengal basin is more seriously affected. This basin covers the major region of states such as West Bengal, Assam, Bihar, and Tripura in India and roughly the entire region of Bangladesh (Chakraborty et al. 2015). The shallow Holocene aquifer present in this basin is the major culprit behind the nuisance of As contamination in this area, which can be regarded as a nonpoint geogenic As source. This As laden ground water when used for irrigation purposes causes a significant accumulation in the soil that is concomitantly transmitted to plants grown on these soils. Food crops such as rice, pulses, and vegetables grown on these soils are consumed by humans and animals, thus facilitating the entry of As into the food chain (Biswas et al. 2019).

Studies in As contaminated areas have revealed that more than the total content in the soil, its bioavailability is largely governed by the form or oxidation state in which it is present. In nature, it occurs in various oxidation states that range from highly reduced −3 state to highly oxidized +5 state depending on pH and redox potential. These are As^{3-} (arsine), As^0 (elemental arsenic), As^{3+} (arsenite) to As^{5+} (arsenate) (Al-Makishah et al. 2020). These species remain in inorganic or organic form. Commonly occurring organic forms of As include monomethyl arsenic acid (MMA) and dimethyl arsenic acid (DMA), while inorganic forms encompass arsenite and arsenate (Fig. 10.1). As^{5+} is a phosphate homologue that can be hazardous if it enters the path of vital phosphate-dependent functions such as ATP generation. Phosphate-carrying proteins can transport As^{5+} into the cell through the cellular membrane, resulting in abnormalities in phosphate supply. As^{3+}, on the other hand, proved to be hazardous due to its tendency to bond with thiols and protein sulfhydryl group and interference with several enzyme processes resulting in cellular damage (Wei et al. 2015). Therefore, the overall ascending order of increasing toxicity of As species varies as follows. DMA < MMA < As^{5+} < As^{3+} (Kamal and Miah 2022). Regardless of the chemical species of As, interaction with As in cells frequently results in the liberation of reactive oxygen species (ROS), which may result in the

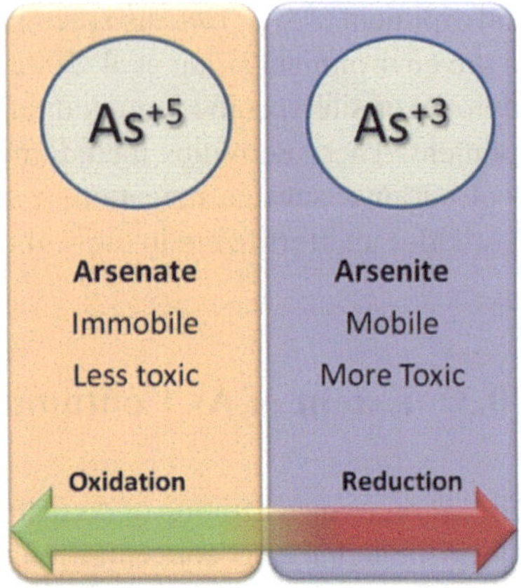

Fig. 10.1 Relative toxicity of inorganic As species

synthesis of several enzymes associated with antioxidant defense. As exposure not only affects the oxidative metabolism of carbon but also impedes the routes for the absorption of nitrogen and sulfur and ultimately protein metabolism (Coelho et al. 2020).

10.4 Remediation of As Contamination

Due to the variety of the detrimental effects of As on plants and humans, scientists around the globe to discover means by which this menace can be controlled. Remediation methods may involve two basic approaches: either these methods eliminate As from polluted media or methods that diminish its biotoxicity (Fig. 10.2). Both these approaches can be designed as suitable remediation options. Currently, we have an array of remediation practices that may fall into either of the categories of—physical, chemical, biological, or integrated methods of As remediation. Physical techniques include soil cover, soil turnover, and electrokinetic remediation. Chemical methods consist of soil washing and immobilization. Biological methods include bioremediation and phytoextraction. Combining one technique with another to supplement the removal efficiency of each technique falls under the category of integrated method of As remediation. For instance, the addition of certain chemicals to increase bioavailability followed immediately by growing crops for phytoextraction to increase overall removal efficiency falls under the integrated method of As remediation (Wan et al. 2019). Different techniques of remediation of As contamination are presented in Fig. 10.2.

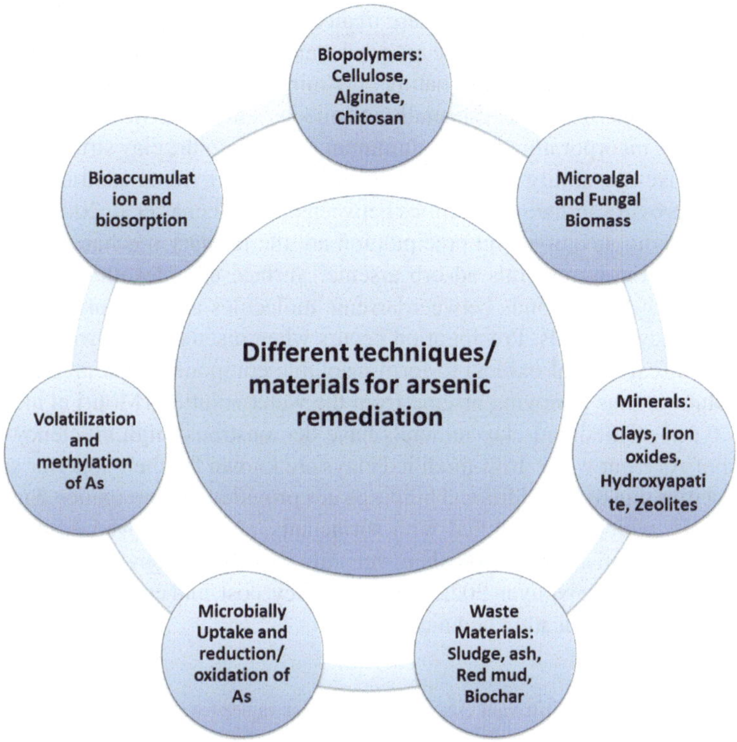

Fig. 10.2 Different techniques/materials for arsenic remediation

10.4.1 Clay Modification Techniques for As Remediation

Clay minerals, characterized by their ion exchange capacities, and high surface area have emerged as good materials for the remediation of arsenic-polluted water. Modification of these clays enhances their adsorption capacity and making them more effective in removing arsenic from polluted water (Meena[a] et al. 2023). Various techniques have been developed to modify clay minerals to increase the adsorption efficiency and stability of the materials (Meena[b] et al. 2023; Mukhopadhyay et al. 2017; Naveenkumar et al. 2023) (Table 10.1). Physical modifications of clay involve processes like thermal treatment, particle size reduction, and mechanical mixing. Thermal treatment changes the structure of the clay, increasing its surface area and porosity, which increases its adsorption capacity (Anyukhina and Seredin 2022). Reducing the particle size increases the surface area-to-volume ratio, providing more active sites for arsenic adsorption on the clay surface. Mechanical mixing, on a other hand, ensures the uniform distribution of clay particles in the solution, facilitating improved contact with arsenic molecules (Pajak 2021). Chemical modifications are more complex and involve treatments with acids, bases, surfactants, and metal oxides (Meena[c] et al. 2023; Naveenkumar et al. 2023). Acid treatment is a common method for arsenic remediation, where clays are treated with acids such as hydrochloric or sulfuric acid to increase their surface area and create more active sites for adsorption. Surfactant modification involves coating clay minerals with surfactants to enhance their ability to adsorb arsenic, especially in aqueous solutions. Metal oxide impregnation, such as incorporating iron or aluminum oxides into the clay structure, significantly improves the ability of clay to adsorb arsenic by providing additional binding sites and improving surface interactions between both (Meena[b] et al. 2023). Adsorption, surface complexation, and precipitation are the primary mechanisms through which modified clay minerals adsorb arsenic. Surface complexation involves the formation of chemical bonds between arsenic molecules and the surface of active sites on the clay minerals. Precipitation occurs when arsenic reacts with the modifying agents (e.g., metal oxides) to form insoluble compounds that precipitate out of the solution, thus removing arsenic from the water solution (Mojiri et al. 2024). Several types of modified clay minerals have demonstrated high efficiency in the removal of As from water. Iron-modified clays are known for their superior adsorption capacities due to the additional binding sites provided by iron oxides. Similarly, organo-clays, which are modified with surfactants, show improved adsorption in various water chemistries, making them versatile for different environmental conditions (Hao et al. 2018; Nguyen 2022). The efficiency, cost, and environmental impact of different clay modification techniques vary.

Table 10.1 Comparison of different clay modification techniques for As removal

Criteria	Metal-impregnated clays	Physical modifications	Chemical modifications
Efficiency	High adsorption capacities	Moderate adsorption capacities	High, depending on the chemicals used (Naveenkumar et al. 2023)
Cost	Higher due to metal incorporation	Lower, cost-effective	Moderate to high, depending on chemicals and processes used (Naveenkumar et al. 2023)
Environmental impact	Potential introduction of secondary contaminants	Low, minimal environmental impact	Potential for introducing secondary contaminants, requires careful handling (Meena[c] et al. 2023)

10.4.2 Microbial Remediation Techniques for As Removal

Elevated As levels disturb beneficial soil microbes, influence plant physiology, and modify soil fertility and characteristics, all of which have an adverse effect on plant growth and development. According to Yang and Rosen (2016), a wide variety of microorganisms are found in soil and are essential to the biogeochemical cycling of As. These processes include bioaccumulation and biotransformation. For example, As may bind to microalgae surfaces and eventually pass through the cell membrane. According to Duncan et al. (2015), As is metabolized by marine algae through methylation, reduction, and oxidation processes and conversion into arseno-sugars. Under different pH and redox circumstances, bacteria, fungi, and algae may convert pentavalent As^{5+} into its trivalent As^{3+} form or methylate it into organic molecules. A summary of current research on the metabolic processes involved in bioremediation, such as absorption, oxidation–reduction reactions, biosorption, methylation, and volatilization, is provided in the following section.

10.4.2.1 Uptake and Reduction/Oxidation of As

According to Mohsin et al. (2023) (Table 10.2), microorganisms use two primary processes to reduce As: external reduction in the periplasmic region and intracellular reduction within the cytoplasm. Phosphate transporters aid in the absorption of arsenate (As(V)) in organisms, and the uptake of As(V) is influenced by phosphate shortage (Preetha et al. 2023). As(V) absorption in *Escherichia coli* is mediated by two different transport mechanisms. According to Zhang et al. (2022), the Pit system constantly functions with moderate affinity and high capacity for As(V) absorption, whereas the Pst system has restricted capacity but a high affinity. Phosphate transport depends on both the Pit and Pho87p membrane proteins (Preetha et al. 2023). As(V) initiates the cytoplasmic reduction pathway by activating the ars operon upon entry into the cell via the Pit and Pst transporters. The Ars-A enzyme uses the electrochemical energy to produce ATP throughout this process. Because As(III) is more susceptible to oxidation than As(V), As oxidation, which is mainly aided by ambient chemicals and microorganisms, is thought to be an efficient bioremediation technique. To investigate its function in oxidizing As(III) to As(V), the enzyme As(III) oxidase (AioBA) was isolated and studied in the bacteria *Alcaligenes faecalis*. AioBA is synthesized by many bacterial species. Furthermore, dissimilatory reduction during anaerobic respiration can transform arsenate into As(III) (Abbas et al. 2018). In research by Wang et al. (2021), the marine microalga *Dunaliella salina* showed that it could oxidize As(III) at a concentration of 5.4 mg/L.

Table 10.2 Microorganisms with the ability to remediate As in soil

Microorganism	Mechanism of bioremediation	Refs.
Bacteria		
Alcaligenes faecalis	Oxidation of A (III)	Irshad et al. (2020) Bagade et al. (2020) Mushtaq et al. (2020)
Acidithiobacillus ferrooxidans	Increased As leaching using mixed microbial cultures	
Acidithiobacillus thiooxidans	Increased As leaching using mixed microbial cultures	
Bacillus flexus strain As-12	Effective remediation of As contamination under laboratory conditions	
Microvirga sp. S-MI1b	Oxidation of As(III)	
Kocuria flava AB402	Efficient adsorption and accumulation of As in As(III) and high-salinity conditions	
Bacillus vietnamesis AB403	Efficient adsorption and accumulation of As in As(III) and high-salinity conditions	
Ancylobacter sp. TS-1 AK1 (KY569423)	Oxidation of As(III)	
Ancylobacter sp. TS-1 AK9 (KY569424)	Oxidation of As(III)	
Bacillus sp. XZM	Enhanced As uptake and removal efficiency with Vallisneria denseserrulata	
Bacillus firmus L-148	Oxidation of As(III)	
Bacillus subtilis S4	Bacillus subtilis S4 with iron oxide nanoparticles enhanced the growth of Cucurbita moschata seedlings under As stress	
Bacillus flexus strain SSAI1	Oxidation of As(III)	Mujawar et al. (2021)
Fungus		
Fusarium oxysporum, *Trichoderma asperellum*, *Penicillium janthinellum*	Demethylation and volatilization	Button et al. (2020) Farkas et al. (2020)
Trichoderma sp. MG	Precipitation of calcite	
Fomitopsis betulina	Accumulation of different As species	
Aspergillus niger strain CBS 140837	Decreased bioavailability of As(V) through manganese oxide biotransformation	
Arbuscular mycorrhizal fungi (AM) fungi	AM fungi inoculation notably reduced As concentration in *Pisum sativum* grains	
Aspergillus niger strain	Extraction of heavy metals, including As	
Rhizophagus intraradices	Bio-amelioration of As-induced physiological disturbances in wheat	
BGA/algae		
Nostoc sp. PCC 7120	Transformation of As	Zhu et al. (2020)
Dunaliella salina	Reduction in uptake of As(III) or As(V)	

(continued)

Table 10.2 (continued)

Microorganism	Mechanism of bioremediation	Refs.
Leptolyngbya boryana	Transformation of As	
Chlorella sp.	Alleviated As toxicity in rice seedlings by altering gene expression related to antioxidant enzymes and methylation	Awasthi et al. (2021)

10.4.2.2 Volatilization and Methylation of As

Microorganisms use a process called As methylation, which is generally acknowledged as a detoxifying mechanism, to convert inorganic As into organic forms. Through the use of enzymes, poisonous methylated As(III) is transformed into methylated As(V), which is less dangerous (Chen and Rosen 2020). This process is aided by the microbial enzyme arsenite S-adenosylmethionine methyltransferase (SAM), which is encoded by arsM genes (Irshad et al. 2021). Trimethylarsine, which is more volatile and less harmful in the environment, is the final product of As(V)'s multistage methylation process. As(III), methylarsonate, dimethylarsinate, dimethylarsinite, and trimethylarsine oxide—which is subsequently reduced to trimethylarsine—are among the intermediates in this process (Stýblo et al. 2021). As methylation in species such as *Rhodopseudomonas palustris* (Yang et al. 2020), *Streptomyces* sp., and *Methanosarcina acetivorans* C2A has been well investigated. Although they yield distinct chemicals, both bacteria and fungi are capable of methylating As. According to Jiang et al. (2022), fungi mostly make trimethylarsine, while bacteria usually produce methylated mercaptans and dimethylarsine.

10.4.2.3 Bioaccumulation and Biosorption

It is well-known that using microbes for biosorption and bioaccumulation is a very efficient way of bioremediating harmful metals and metalloids. Energy-dependent mechanisms that include chelation, ion exchange, and physical absorption that are made possible by physicochemical interactions are involved in the microbial biosorption of aspartame (Preetha et al. 2023). Cellulose and lignin are typical biological sorbents used in this procedure (Wu et al. 2021). By interacting with As via several functional groups as –COOH, –SH, –OH, $-NH_2$, and $-PO_3$, microorganisms improve biosorption (Yin et al. 2022; Simon et al. 2024). On the other hand, the process by which organisms gather and store contaminants within their biomass is known as bioaccumulation. Microorganisms use phosphate transporters and glycerol to sequester free As during bioaccumulation (Mohsin et al. 2023). When it is attached to proteins or peptides, As is able to reduce its toxicity via this process.

10.4.3 Arsenic Remediation Using Genetically Modified Microorganisms

Strategies for biotechnological removal of As have focused on understanding of certain aspects of molecular biology regulating arsenic uptake, metabolism, and translocation in organisms. Furthermore, the microbial bioremediation techniques described so far follow the conventional use of microbes, capable of being genetically modified to improve As accumulation and subsequent removal. This is achieved by modification of microbes aiming to regulate genes such as *ArsC* to reduce arsenate reduction, modification of As transporters such as *ArsB* and *Acr3* aimed at improving As efflux and addition of genes that allow As methylation of As into less toxic forms. The transgenic microbes are being studied for use in water treatments, soil, and industrial effluent waste disposal systems. Table 10.3 represents genetically modified microorganisms that are engineered for As bioremediation.

10.4.4 Transgenic Plants in As Removal

Various transgenic plants have been developed that can accumulate or detoxify As from contaminated environments. For the last few decades, genetic engineering related to As resistance and detoxification mechanisms has been highly researched under different conditions. The use of different approaches was applied to enhance As resistance and reduce its accumulation in the plants. The overexpression of the *WaarsM* gene in rice resulted in high As tolerance and lowered its concentration in the grains through enhanced As volatilization. Similarly, overexpression of *PvPHT1* in *Pteris vittata* results in higher uptake because of its higher affinity (Sun et al. 2019). *The expression of PvACR3* in *Arabidopsis thaliana* athac1 mutant transformed the plant into a hyperaccumulator of As. This technique can then be used for some fast-growing and high-biomass crop species to efficiently perform phytoremediation. Genetic transformation of the *AtACR2* gene from *A. thaliana* substantially decreased As accumulation in tobacco shoots (*Nicotiana tabacum* var. Sumsun) (Nahar et al. 2017). The *ArsC* gene obtained from the fern *Pityrogramma calomelanos* improved As tolerance of As in the transgenic lines. Recent studies have identified that the *OsARM1* MYB transcription factor gene is an important transcriptional regulator in the response to As in rice. The gene can be strongly induced by As(III) and negatively regulates the expression of As-associated transporter genes, including *OsLsi1*, *OsLsi2*, and *OsLsi6*. In rice, mutation in the As transporter genes *OsPht1:8* and *Lsi1/2* resulted in reduced accumulation of As in grains. Meanwhile, overexpressions of mutations in the *OsPT8* gene in two rice varieties, Nipponbare and Kasalath, showed a reduction in As(V) uptake of 33–57% and an enhanced As(V) tolerance to As(V) compared to wild-type plants. The results indicate the amenability of genetic traits to manipulation with the aim of improving As phytoremediation for environmental benefits using transgenic approaches or the expression of particular genes.

Table 10.3 Genetically modified microorganisms engineered for arsenic bioremediation

Engineered microorganism	Targeted gene	Source organism	Arsenic removal	Mechanism involved	Refs.
Sphingomonas desiccabilis and *Bacillus idriensis*	*ArsM*	*Rhodopseudomonas palustris*	2.92 ± 0.04 mg kg^{-1} As(III)	Biovolatilization	Liu et al. (2011)
Saccharomyces cerevisiae	*PHO84*	*Saccharomyces cerevisiae*	2,600–2,800 μg kg^{-1} As(V)	Immmobilization	Shen et al. (2012)
Escherichia coli BL21(DE3)	*ArsR*	Metagenomically mined	0.95 ± 0.02 mg kg^{-1} As(III)	Bioaccumulation	Yang et al. (2013)
Pseudomonas putida KT2440	*ArsM*	*Chlamydomonas reinhardtii*	57% of total As	Methylation	Chen et al. (2013)
Bacillus subtilis 168	*CmarsM*	*Cyanidioschyzon merolae*	$78.77 \pm 1.33\%$ of the total As	Biovolatilization	Huang et al. (2015)
Escherichia coli Rosetta (DE3)	*OsMT*	*Oryzae sativa*	1.36 ± 0.39 mg kg^{-1} As(III)	Bioaccumulation	Shahpiri and Mohammadzadeh (2018)
Escherichia coli	*ArsRRP1–4*	*Rhodopseudomonas palustris* CGA009	2.32 mg g^{-1} As(III) and 1.47 mg g^{-1} As(V)	Adsorption	Ke et al. (2018)
Saccharomyces cerevisiae	*WaarsM*	*Westerdykella aurantiaca*	–	Biovolatilization	Verma et al. (2019)
Chlamydomonas reinhardtii CC-125 (mtþ) c137	*Acr3*	*Pteris vittata*	–	Biotransformation	Ramírez-Rodríguez et al. (2019)
Escherichia coli Rosetta (DE3) and *Escherichia coli* Rosetta gami 2	*CgArsR1*	*Corynebacterium glutamicum*	1200 μg g^{-1} dry cell weight (DCW) As(III) and 338 μg g^{-1} DCW As(V)	Bioaccumulation	Maleki and Shahpiri (2022)
Pseudomonas putida KT2440	*ArsR*	Metagenomically mined	$92.08 \pm 1.23\%$ of total As	Adsorption	Xue et al. (2017)

The discovery of such transcriptional regulators is the pathway to the development of genetically modified transgenic crops with the ability to remediate As.

10.5 Conclusions

The management of As contamination in soils requires a multifaceted approach, incorporating innovative technologies and remediation strategies. Several remediation approaches, including microbial treatment and clay modification, show promise, although each of these remediation methods has limitations and challenges that need to be further developed and improved. Knowing the processes that lead to As absorption, oxidation, reduction, volatilization, and methylation is critical to better and improved remediation strategies. Additionally, bioaccumulation and biosorption functions also provide environmentally acceptable and sustainable solutions to the elimination of As. Most importantly, going forward, these many methods will need to be brought into coherent and useful applications.

Acknowledgements The authors sincerely thank our prestigious institute and other people who are involved in this study for their direct and indirect contributions.

References

Abbas, G., Murtaza, B., Bibi, I., Shahid, M., Niazi, N. K., Khan, M. I., and Natasha, 2018. Arsenic uptake, toxicity, detoxification, and speciation in plants: physiological, biochemical, and molecular aspects. *International Journal of Environmental Research and Public Health*, 15(1), 59.

Ahmed, S. F., Kumar, P. S., Rozbu, M. R., Chowdhury, A. T., Nuzhat, S., Rafa, N., Mahlia, T. M. I., Ong, H. C., and Mofijur, M, 2022. Heavy metal toxicity, sources, and remediation techniques for contaminated water and soil. Environmental Technology & Innovation, 25, 102114. https://doi.org/10.1016/j.eti.2021.102114

Al-Makishah, N. H., Taleb, Md. A., and Barakat, M. A. 2020. Arsenic bioaccumulation in arsenic-contaminated soil: A review. Chemical Papers, 74(9), 2743–2757. https://doi.org/10.1007/s11696-020-01122-4

Anyukhina, A. V., and Seredin, V. V. 2022. Concerning Adsorption of Clays Subjected by Heat and Chemical Activation. In A. Rocha and E. Isaeva (Eds.), Science and Global Challenges of the 21st Century—Science and Technology (Vol. 342, pp. 218–226). Springer International Publishing. https://doi.org/10.1007/978-3-030-89477-1_21

Awasthi, S., Chauhan, R., Dwivedi, S., Srivastava, S., Srivastava, S., and Tripathi, R. D. 2018. A consortium of alga (*Chlorella vulgaris*) and bacterium (*Pseudomonas putida*) for amelioration of arsenic toxicity in rice: A promising and feasible approach. *Environmental and Experimental Botany*, 150, 115–126.

Awasthi, S., Chauhan, R., Indoliya, Y., Chauhan, A. S., Mishra, S., Agrawal, L., and Tripathi, R. D. 2021. Microbial consortium mediated growth promotion and arsenic reduction in rice: an integrated transcriptome and proteome profiling. *Ecotoxicology and Environmental Safety*, 228, 113004.

Bagade, A., Nandre, V., Paul, D., Patil, Y., Sharma, N., Giri, A., and Kodam, K. 2020. Characterisation of hyper tolerant Bacillus firmus L-148 for arsenic oxidation. *Environmental Pollution,* 261, 114124.

Biswas, A., Swain, S., Chowdhury, N. R., Joardar, M., Das, A., Mukherjee, M., and Roychowdhury, T. 2019. Arsenic contamination in Kolkata metropolitan city: Perspective of transportation of agricultural products from arsenic-endemic areas. *Environmental Science and Pollution Research*, 26(22), 22929–22944. https://doi.org/10.1007/s11356-019-05595-z

Button, M., Koch, I., Watts, M. J., and Reimer, K. J. 2020. Arsenic speciation in the bracket fungus *Fomitopsis betulina* from contaminated and pristine sites. *Environmental Geochemistry and Health*, 42, 2723–2732.

Chakraborty, M., Mukherjee, A., and Ahmed, K. M. 2015. A Review of Groundwater Arsenic in the Bengal Basin, Bangladesh and India: From Source to Sink. Current Pollution Reports, 1(4), 220–247. https://doi.org/10.1007/s40726-015-0022-0

Chen, J., and Rosen, B. P. 2020. The arsenic methylation cycle: how microbial communities adapted methylarsenicals for use as weapons in the continuing war for dominance. *Frontiers in Environmental Science,* 8, 43.

Chen, J., Qin, J., Zhu, Y.-G., de Lorenzo, V., and Rosen, B. P. 2013. Engineering the Soil Bacterium *Pseudomonas putida* for Arsenic Methylation. *Applied and Environmental Microbiology*, 79(14), 4493–4495.

Coelho, D. G., Marinato, C. S., de Matos, L. P., de Andrade, H. M., da Silva, V. M., Santos-Neves, P. H., Araújo, S. C., and Oliveira, J. A. 2020. Is arsenite more toxic than arsenate in plants? *Ecotoxicology*, 29(2), 196–202. https://doi.org/10.1007/s10646-019-02152-9

Duncan, E. G., Maher, W. A., and Foster, S. D. 2015. Contribution of arsenic species in unicellular algae to the cycling of arsenic in marine ecosystems. *Environmental Science & Technology*, 49(1), 33–50.

Farkas, B., Kolenčík, M., Hain, M., Dobročka, E., Kratošová, G., Bujdoš, M., and Urík, M. 2020. *Aspergillus niger* decreases bioavailability of arsenic (V) via biotransformation of manganese oxide into biogenic oxalate minerals. *Journal of Fungi,* 6(4), 270.

Finnegan, P. M., and Chen, W. 2012. Arsenic Toxicity: The Effects on Plant Metabolism. *Frontiers in Physiology*, 3, 182. https://doi.org/10.3389/fphys.2012.00182

Hao, L., Liu, M., Wang, N., and Li, G. 2018. A critical review on arsenic removal from water using iron-based adsorbents. RSC Advances, 8(69), 39545–39560.

Huang, K., Chen, C., Shen, Q., Rosen, B. P., and Zhao, F. J. 2015. Genetically engineering *Bacillus subtilis* with a heat-resistant arsenite methyltransferase for bioremediation of arsenic-contaminated organic waste. *Applied and environmental microbiology*, 81(19), 6718–6724.

Irshad, S., Xie, Z., Wang, J., Nawaz, A., Luo, Y., Wang, Y., and Mehmood, S. 2020. Indigenous strain *Bacillus* XZM assisted phytoremediation and detoxification of arsenic in Vallisneria denseserrulata. *Journal of hazardous materials*, 381, 120903.

Irshad, S., Xie, Z., Mehmood, S., Nawaz, A., Ditta, A., and Mahmood, Q. 2021. Insights into conventional and recent technologies for arsenic bioremediation: A systematic review. *Environmental Science and Pollution Research*, 28, 18870–18892.

Jiang, Z., Shen, X., Shi, B., Cui, M., Wang, Y., and Li, P. 2022. Arsenic mobilization and transformation by ammonium-generating bacteria isolated from high arsenic groundwater in Hetao Plain, China. *International Journal of Environmental Research and Public Health*, 19(15), 9606.

Kamal, M. Z. U., and Miah, M. Y. 2022. Arsenic speciation techniques in soil water and plant: An overview. Arsenic Monitoring, Removal and Remediation; IntechOpen: London, UK, 9.

Ke, C., Zhao, C., Rensing, C., Yang, S., and Zhang, Y. 2018. Characterization of recombinant *E. coli* expressing arsR from *Rhodopseudomonas palustris* CGA009 that displays highly selective arsenic adsorption. *Applied Microbiology and Biotechnology*, 102(14), 6247–6255.

Liao, Z.-H., Chuang, H.-C., Huang, H.-T., Wang, P.-H., Chen, B.-Y., Lee, P.-T., Wu, Y.-S., and Nan, F.-H. 2022. Bioaccumulation of arsenic and immunotoxic effect in white shrimp (*Penaeus vannamei*) exposed to trivalent arsenic. *Fish & Shellfish Immunology*, 122, 376–385.

Liu, S., Zhang, F., Chen, J., and Sun, G. 2011. Arsenic removal from contaminated soil via biovolatilization by genetically engineered bacteria under laboratory conditions. *Journal of Environmental Sciences*, 23(9), 1544–1550.

Maleki, F., and Shahpiri, A. 2022. Efficient and specific bioaccumulation of arsenic in the transgenic *Escherichia coli* expressing ArsR1 from *Corynebacterium glutamicum. Biometals*, 35(5), 889–901.

Meena[a], S., Chobhe, K. A., Golui, D., Bana, R. S., Gawdiya, S., and Roy, A. 2023. Assessing the impact of altered clay mineral and red mud derivatives on the characteristics of mustard (*B. juncea*) and the Soil Arsenic Content. *International Journal of Plant & Soil Science*, 35(19), 200–206.

Meena[b], S., Chobhe, K. A., Manjaiah, K. M., Datta, S. P., and Golui, D. 2023. Immobilization of arsenic in soil using modified bentonite and red mud to reduce its bio-availability in *Brassica juncea. The Indian Journal of Agricultural Sciences*, 93(9), 1049–1052.

Meena[c], S., Chobhe, K. A., Manjaiah, K. M., Datta, S. P., Golui, D., Raza, M. B., Rahman, M. M., and Naveenkumar, A. 2023. Synthesis, characterization and adsorptive performances of functionalized clay minerals and red mud for aqueous arsenic removal. *Groundwater for Sustainable Development*, 23, 101025.

Mohsin, H., Shafique, M., Zaid, M., and Rehman, Y. 2023. Microbial biochemical pathways of arsenic biotransformation and their application for bioremediation. *Folia Microbiologica*, 68(4), 507–535.

Mojiri, A., Razmi, E., KarimiDermani, B., Rezania, S., Kasmuri, N., Vakili, M., and Farraji, H. 2024. Adsorption methods for arsenic removal in water bodies: A critical evaluation of effectiveness and limitations. *Frontiers in Water*, 6, 1301648.

Mujawar, S. Y., Vaigankar, D. C., and Dubey, S. K. 2021. Biological characterization of *Bacillus flexus* strain SSAI1 transforming highly toxic arsenite to less toxic arsenate mediated by periplasmic arsenite oxidase enzyme encoded by aioAB genes. *Bio-Metals*, 34(4), 895–907.

Mukhopadhyay, R., Manjaiah, K. M., Datta, S. C., Yadav, R. K., and Sarkar, B. 2017. Inorganically modified clay minerals: Preparation, characterization, and arsenic adsorption in contaminated water and soil. *Applied Clay Science*, 147, 1–10.

Mushtaq, T., Shah, A. A., Akram, W., and Yasin, N. A. 2020. Synergistic ameliorative effect of iron oxide nanoparticles and *Bacillus subtilis* S4 against arsenic toxicity in *Cucurbita moschata*: polyamines, antioxidants, and physiochemical studies. *International journal of phytoremediation*, 22(13), 1408–1419.

Nahar, N., Rahman, A., Ghosh, S., Nawani, N., and Mandal, A. 2017. Functional studies of AtACR2 gene putatively involved in accumulation, reduction and/or sequestration of arsenic species in plants. *Biologia*, 72(5), 520–526.

Naveenkumar, A., Manjaiah, K. M., Sharma, V. K., Ray, P., Pandey, P., Sahoo, R. N., Kumar, R., Das, T. K., Raza, M. B., and Meena, S. 2023. Assessing the impact of functionalized bentonite in reducing lead content and uptake in Spinach (*Spinacia oleracea* cv. All Green). Biological *Forum–An International Journal*, 15(8), 135–140.

Nguyen, T. H. 2022. Arsenic Removal from Water by Novel Iron-Based Adsorbents. University of Technology Sydney (Australia).

Pajak, M. 2021. Adsorption Capacity of Smectite Clay and Its Thermal and Chemical Modification for Two Anionic Dyes: Comparative Study. *Water, Air, & Soil Pollution*, 232(2), 83. https://doi.org/10.1007/s11270-021-05032-3

Podgorski, J., and Berg, M. 2020. Global threat of arsenic in groundwater. *Science*, 368(6493), 845–850. https://doi.org/10.1126/science.aba1510

Preetha, J. S. Y., Arun, M., Vidya, N., Kowsalya, K., Halka, J., and Ondrasek, G. 2023. Biotechnology advances in bioremediation of arsenic: a review. *Molecules*, 28(3), 1474.

Ramírez-Rodríguez, A.E., Bañuelos-Hernández, B., García-Soto, M.J., Govea-Alonso, D.G., Rosales-Mendoza, S., Alfaro de la Torre, M.C., Monreal-Escalante, E. and Paz-Maldonado, L.M., 2019. Arsenic removal using *Chlamydomonas reinhardtii* modified with the gene acr3 and enhancement of its performance by decreasing phosphate in the growing media. *International journal of phytoremediation*, 21(7), pp.617–623.

Sarkar, A., Paul, B., and Darbha, G. K. 2022. The groundwater arsenic contamination in the Bengal Basin-A review in brief. *Chemosphere*, 299, 134369.

Shahpiri, A., and Mohammadzadeh, A. 2018. Bioaccumulation of arsenic by engineered *Escherichia coli* cells expressing rice metallothionein isoforms. *Current microbiology*, 75, 1537–1542.

Shen, M. W. Y., Shah, D., Chen, W., and Da Silva, N. 2012. Enhanced arsenate uptake in *Saccharomyces cerevisiae* overexpressing the Pho84 phosphate transporter. *Biotechnology Progress*, 28(3), 654–661.

Shirin, S., & Yadav, A. K. (2021). Assessment and suitability for the dumping of flyash in an abandoned mine. Environmental Challenges, 4, 100111. https://doi.org/10.1016/j.envc.2021.100111

Simon, M., Joshi, H., Yadav, A. K., & Hopke, P. K. (2024). Simultaneous nitrogen and phosphorus removal from polluted water using a novel bacterial consortium (BACON-38): Promising application for wastewater remediation. Journal of Water Process Engineering, 66, 106044. https://doi.org/10.1016/j.jwpe.2024.106044

Stýblo, M., Venkatratnam, A., Fry, R. C., and Thomas, D. J. 2021. Origins, fate, and actions of methylated trivalent metabolites of inorganic arsenic: progress and prospects. *Archives of toxicology*, 95, 1547–1572.

Sun, D., Feng, H., Li, X., Ai, H., Sun, S., Chen, Y., Xu, G., Rathinasabapathi, B., Cao, Y. and Ma, L.Q. 2019. Expression of new *Pteris vittata* phosphate transporter PvPht1; 4 reduces arsenic translocation from the roots to shoots in tobacco plants. *Environmental science and technology*, 54(2), 1045–1053.

Verma, S., Verma, P. K., and Chakrabarty, D. 2019. Arsenic bio-volatilization by engineered yeast promotes rice growth and reduces arsenic accumulation in grains. *International Journal of Environmental Research*, 13, 475–485.

Wan, X., Lei, M., and Chen, T. 2019. Review on remediation technologies for arsenic-contaminated soil. *Frontiers of Environmental Science & Engineering*, 14(2), 24. https://doi.org/10.1007/s11783-019-1203-7

Wang, Y., Wang, S., Xu, P., Liu, C., Liu, M., Wang, Y., and Ge, Y. 2015. Review of arsenic speciation, toxicity and metabolism in microalgae. *Reviews in Environmental Science and Bio/Technology*, 14, 427–451.

Wang, Y., Li, J., Ma, T., Xie, X., Deng, Y. and Gan, Y., 2021. Genesis of geogenic contaminated groundwater: As, F and I. *Critical Reviews in Environmental Science and Technology*, *51*(24), pp. 2895–2933.

Wei, C., Ge, Z., Chu, W., and Feng, R. 2015. Speciation of antimony and arsenic in the soils and plants in an old antimony mine. *Environmental and Experimental Botany*, 109, 31–39. https://doi.org/10.1016/j.envexpbot.2014.08.002

World Health Organization. 2022. *Guidelines for drinking-water quality: incorporating the first and second addenda*. World Health Organization.

Wu, Q., Ren, M., Zhang, X., Li, C., Li, T., Yang, Z., and Wang, L. 2021. Comparison of Cd (II) adsorption properties onto cellulose, hemicellulose and lignin extracted from rice bran. Lwt, 144, 111230.

Xue, X. M., Yan, Y., Xiong, C., Raber, G., Francesconi, K., Pan, T., and Zhu, Y. G. 2017. Arsenic biotransformation by a cyanobacterium *Nostoc sp.* PCC 7120. *Environmental pollution*, 228, 111–117.

Yadav, A. K., Sahoo, S. K., Mahapatra, S., Kumar, A. V., Pandey, G., Lenka, P., & Tripathi, R. M. (2014). Concentrations of uranium in drinking water and cumulative, age-dependent radiation doses in four districts of Uttar Pradesh, India. Toxicological & Environmental Chemistry, 96(2), 192–200. https://doi.org/10.1080/02772248.2014.934247

Yang, H. C., and Rosen, B. P. 2016. New mechanisms of bacterial arsenic resistance. *Biomedical journal*, 39(1), 5–13.

Yang, T., Liu, J.-W., Gu, C., Chen, M.-L., and Wang, J.-H. 2013. Expression of Arsenic Regulatory Protein in *Escherichia coli* for Selective Accumulation of Methylated Arsenic Species. *ACS Applied Materials and Interfaces*, 5(7), 2767–2772.

Yang, P., Ke, C., Zhao, C., Liu, B., Xue, X., Rensing, C., and Yang, S. 2020. ArsM-mediated arsenite volatilization is limited by efflux catalyzed by As efflux transporters. *Chemosphere*, 239, 124822.

Yin, S., Zhang, X., Yin, H., and Zhang, X. 2022. Current knowledge on molecular mechanisms of microorganism-mediated bioremediation for arsenic contamination: A review. *Microbiological Research*, 258, 126990.

Zhang, J., Liu, J., Zheng, F., Yu, M., Shabala, S., and Song, W. Y. 2022. Comparative analysis of arsenic transport and tolerance mechanisms: Evolution from prokaryote to higher plants. *Cells*, 11(17), 2741.

Zhu, F., Yang, M., Luo, Z. X., Yu, R. L., Hu, G. R., and Yan, Y. 2020. Bioaccumulation and biotransformation of arsenic in *Leptolyngbya boryana*. *Environmental Science and Pollution Research*, 27, 29993–30000.

Chapter 11
Overview of Conventional Arsenic Removal Methods: Strengths and Limitations

Mohd Akhter Ali and M. Kamraju

Abstract Arsenic contamination in drinking water is a significant global health concern, affecting millions of people, particularly in regions dependent on groundwater sources. Effective remediation requires a comprehensive understanding and application of arsenic removal technologies. This chapter provides an extensive review of conventional arsenic removal methods, focusing on their mechanisms, operational principles, practical applications, strengths, and limitations. Methods covered include coagulation-flocculation, adsorption, ion exchange, membrane filtration, and oxidation. Coagulation-flocculation involves the addition of coagulants to form flocs that adsorb arsenic, followed by sedimentation and filtration, with a discussion of the types of coagulants used and their effectiveness under varying water chemistries. Adsorption techniques analyse various adsorbents, such as activated alumina, iron-based adsorbents, and natural materials, evaluating factors such as surface area, porosity, and regeneration capabilities. Ion exchange explores the use of resins to selectively remove arsenic ions, assessing the performance of different resins, regeneration processes, and the impact of competing ions. Membrane filtration techniques, including reverse osmosis, nanofiltration, and ultrafiltration, are examined for their high removal efficiencies, balanced against limitations such as membrane fouling and high operational costs. Oxidation processes, including chemical oxidation and advanced oxidation processes (AOPs), are discussed for their role in converting arsenic to a more easily removable form, evaluating the effectiveness of various oxidizing agents, and integration with other treatment methods. Each method is scrutinized for its strengths, including removal efficiency, ease of implementation, and cost-effectiveness, while also considering limitations such as operational complexity, maintenance requirements, and environmental impacts. By providing a detailed analysis of these conventional methods, this chapter aims to provide a comprehensive understanding of their current state and identify areas where further research and

M. A. Ali (✉)
Department of Geography and Joint Director, Directorate of Admissions, Osmania University, Hyderabad, India
e-mail: drmohdakhterali@gmail.com

M. Kamraju
Faculty of Geography, Department of Geography, APSG, Hyderabad, India

A. K. Yadav et al. (eds.), *Arsenic Removal Technologies*,
https://doi.org/10.1007/978-3-031-91171-2_11

innovation are needed, supporting the development of advanced and sustainable arsenic remediation technologies to ensure safe drinking water worldwide.

Keywords Arsenic removal · Coagulation-flocculation · Adsorption · Ion exchange · Membrane filtration · Oxidation · Water treatment · Remediation technologies · Groundwater contamination

11.1 Introduction

Overview of Arsenic Contamination in Drinking Water

Arsenic, a naturally occurring metalloid, is present in the environment and can contaminate groundwater through both natural and anthropogenic activities. Naturally, arsenic contamination occurs through the weathering of arsenic-containing rocks and minerals, volcanic eruptions, and forest fires. Human activities such as mining, the use of arsenic-based pesticides, industrial processes, and improper disposal of arsenic-laden waste can exacerbate this contamination (Smedley and Kinniburgh 2002; Ravenscroft et al. 2009). Arsenic groundwater contamination is a critical issue in many regions around the world, affecting millions of people who rely on these water sources for drinking and irrigation.

In South Asia, particularly in Bangladesh and West Bengal, India, millions of people are exposed to arsenic-contaminated groundwater, with levels often exceeding the World Health Organization (WHO) guideline of 10 μg/L (Chowdhury et al. 2000). In the United States, regions such as the Midwest and parts of the Southwest also report significant arsenic contamination due to both natural and industrial sources (Welch et al. 2000). In Southeast Asia, countries like Vietnam and Cambodia face similar challenges, primarily due to natural geogenic sources (Berg et al. 2001). The widespread nature of arsenic contamination requires a global response to ensure the safety of drinking water.

Arsenic, a toxic metalloid, is widely found in the Earth's crust and enters water sources through both natural and human-induced processes (Smedley and Kinniburgh 2002). Groundwater contamination by arsenic is a serious environmental and public health challenge, particularly for communities dependent on well water. Natural sources of arsenic pollution include the breakdown of arsenic-rich minerals, geothermal activity, and volcanic emissions (Ravenscroft et al. 2009). Meanwhile, industrial operations, mining activities, agricultural pesticides, and fossil fuel combustion contribute to its release into water bodies (Shirin et al. 2023).

The severity of arsenic contamination varies worldwide, with South Asia being one of the most affected regions. In Bangladesh and West Bengal, India, millions consume arsenic-laden groundwater, where concentrations frequently surpass the World Health Organization (WHO) safety limit of 10 μg/L, with some areas recording levels between 300 and 1000 μg/L. This contamination is largely attributed to the dissolution of arsenic-bearing iron oxides within the alluvial aquifers of the

Bengal Delta Plain (Bhattacharya et al. 2002). In Southeast Asia, particularly in Vietnam, Cambodia, and Myanmar, groundwater arsenic pollution stems from natural sedimentary deposits, leading to widespread exposure.

Beyond Asia, arsenic contamination is also prominent in parts of South America, including Argentina, Chile, and Mexico, where volcanic and geothermal activity significantly influence groundwater arsenic levels. Similarly, in the United States, arsenic-contaminated aquifers are found in the Midwest, Southwest, and Northeast, often linked to industrial pollution, geothermal sources, and historical mining operations.

Long-term exposure to arsenic in drinking water has been associated with serious health risks, including cancers (lung, skin, and bladder), cardiovascular disorders, neurological impairments, and developmental issues. Arsenicosis, a condition resulting from prolonged arsenic ingestion, leads to severe systemic toxicity and remains a critical concern for affected populations. Given these health implications, effective arsenic removal strategies are essential. Various treatment technologies such as coagulation-flocculation, adsorption, ion exchange, membrane filtration, and biological methods have been developed to address arsenic contamination. However, each approach presents unique challenges related to cost, efficiency, and long-term sustainability, highlighting the need for continuous advancements in arsenic removal techniques.

11.2 Global Health Impacts of Arsenic Exposure

Chronic exposure to arsenic in drinking water poses severe and widespread health risks. Arsenic is a well-documented carcinogen and prolonged ingestion is linked to various cancers, including skin, lung, bladder, liver, and kidney cancers (Smith et al. 2000; Yadav et al. 2014). Noncancer effects include skin lesions, peripheral neuropathy, gastrointestinal symptoms, diabetes, and cardiovascular disease (Naujokas et al. 2013). Arsenic exposure is also associated with developmental effects in children, such as impaired cognitive development and increased risk of respiratory diseases (Vahter 2008).

Skin manifestations such as hyperpigmentation, hypopigmentation, and keratosis are often the first signs of chronic arsenic poisoning (Guha Mazumder 2008). Cardiovascular impacts, including hypertension and ischemic heart disease, are also significant concerns (Chen et al. 2011). Studies have shown that arsenic exposure can alter endocrine function and glucose metabolism, contributing to the development of diabetes (Paul et al. 2007).

The International Agency for Research on Cancer (IARC) classifies arsenic and arsenic compounds as Group 1 carcinogens, emphasizing their high potential to cause cancer in humans (IARC 2004). The World Health Organization (WHO) recognizes arsenic contamination of drinking water as a major public health issue, urging immediate action to mitigate exposure (World Health Organization 2018).

The widespread nature of arsenic contamination and its severe health impacts underscore the urgent need for effective remediation technologies. Understanding the strengths and limitations of conventional arsenic removal methods is crucial for the development of efficient and sustainable solutions to ensure safe drinking water globally. This review aims to provide a comprehensive overview of these methods, providing insight into their operational principles, effectiveness, and potential areas of improvement.

11.3 Objectives of the Study

Aim to Review and Analyse Conventional Arsenic Removal Methods

The primary objective of this study is to perform an exhaustive review and analysis of conventional arsenic removal methods. Arsenic contamination in drinking water is a global, pervasive issue that poses significant health risks, necessitating the deployment of effective remediation technologies. This study will systematically examine the various conventional methods used to remove arsenic from drinking water, including but not limited to coagulation-flocculation, adsorption, ion exchange, membrane filtration, and oxidation.

Each method will be analysed in terms of its operational principles, including the scientific and technical processes involved in the removal of arsenic. For instance, coagulation-flocculation involves adding coagulants to water to form flocs that can adsorb arsenic particles, which are then removed by sedimentation. Adsorption techniques use materials such as activated alumina or granular ferric hydroxide to capture arsenic ions from water. Ion exchange processes replace arsenic ions in water with less harmful ions using resin-based systems. Membrane filtration methods, such as reverse osmosis and nano-filtration, physically separate arsenic from water using selective barriers. Oxidation processes convert arsenic from a more toxic form (As(III)) to a less toxic and more easily removable form (As(V)).

11.4 Coagulation-Flocculation

Mechanism and Operational Principles

Process Description and Chemical Reactions Involved

Coagulation-flocculation is a fundamental process in water treatment, particularly effective in removing arsenic. This method involves two sequential steps: coagulation, where colloidal particles occur, and flocculation, where these particles agglomerate into larger aggregates that can be removed through sedimentation or filtration.

Coagulation

During the coagulation phase, specific chemicals known as coagulants are introduced to the water. The primary purpose of these coagulants is to neutralize the suspended charges on the particles in the water, including arsenic contaminants. The most commonly used coagulants are aluminum sulfate (alum), ferric chloride, and ferric sulfate. These coagulants work by undergoing hydrolysis reactions that produce hydroxide precipitates. For example, when aluminum sulfate (alum) is added to water, it dissociates into aluminum ions and sulfate ions:

$$Al_2(SO_4)_3 \cdot 18H_2O \rightarrow 2Al^{3+} + 3SO_4^{2-} + 18H_2O$$

The aluminum ions (Al^{3+}) then hydrolyze to form aluminum hydroxide:

$$Al^{3+} + 3H_2O \rightarrow Al(OH)_3 \downarrow +3H^+$$

This aluminum hydroxide precipitate acts as a sweep floc, capturing and binding to the arsenic particles suspended in the water, facilitating their aggregation into larger particles that can settle out or be filtered from the water (Smedley and Kinniburgh 2002).

Similarly, ferric chloride, another common coagulant, dissociates into ferric ions and chloride ions upon addition to water:

$$FeCl_3 \rightarrow Fe^{3+} + 3Cl^-$$

The ferric ions (Fe^{3+}) then undergo hydrolysis to form ferric hydroxide:

$$Fe^{3+} + 3H_2O \rightarrow Fe(OH)_3 \downarrow +3H^+$$

Ferric hydroxide similarly functions as a sweep floc, effectively capturing arsenic particles and aiding their removal from the water (Ravenscroft et al. 2009).

Flocculation

After coagulation, the flocculation stage involves a gentle mixing of the water to promote collisions between the destabilized particles, forming larger aggregates known as flocs. These flocs are more easily removed by sedimentation or filtration. The efficiency of flocculation is influenced by several factors, including mixing intensity, duration, and the presence of natural organic matter in the water, which can enhance or inhibit floc formation (Smith et al. 2000).

11.5 Types of Coagulants and Their Effectiveness

Aluminum-Based Coagulants

Aluminum sulfate (alum) is one of the most widely used coagulants because of its effectiveness and availability. The alum works well on a range of pH levels, typically between 5.5 and 8.0, making it versatile for different types of water. The formed aluminum hydroxide flocs are robust and effective at capturing a variety of contaminants, including arsenic. However, the use of alum can lead to residual aluminum in treated water, which poses health concerns and requires careful control of dosage and pH (Smedley and Kinniburgh 2002).

Ferric-Based Coagulants

Ferric chloride and ferric sulfate are also commonly used coagulants. They are particularly effective in waters with higher pH levels, often above 7.0. Ferric hydroxide flocs are generally denser and settle more quickly than aluminum hydroxide flocs, which can improve the efficiency of the treatment process. Additionally, ferric coagulants are less likely to leave residual iron in treated water at typical dosages, reducing potential health risks (Ravenscroft et al. 2009).

Polymeric Coagulants

In addition to metal-based coagulants, polymeric coagulants such as polyaluminum chloride (PAC) are used. These coagulants offer several advantages, including the formation of stronger and more shear-resistant flocs, which can improve the robustness of the coagulation-flocculation process. Polymeric coagulants also function effectively over a broader pH range and can reduce the total dose of coagulant required, minimizing potential side effects (Smith et al. 2000).

Coagulation-flocculation remains a cornerstone in the treatment of arsenic-contaminated water because of its simplicity and effectiveness. The choice of coagulant and operating parameters must be carefully optimized to maximize arsenic removal while minimizing any negative impacts on water quality. Understanding the chemical reactions and mechanisms involved provides a foundation for improving these processes and developing more effective and sustainable water treatment solutions.

11.6 Strengths of Coagulation-Flocculation

Efficiency in Removing Arsenic at Low Concentrations

Coagulation-flocculation is highly efficient in removing arsenic from water, even at low concentrations. This method effectively targets both soluble arsenic species like arsenite (As(III)\{As(III)}As(III)) and particulate arsenic, converting them into larger, settleable flocs that can be easily removed during subsequent treatment steps.

Studies have demonstrated its ability to reduce arsenic levels to well below regulatory limits, making it suitable for treating water sources with varying levels of arsenic contamination (Smedley and Kinniburgh 2002).

Cost-Effectiveness

From a cost perspective, coagulation-flocculation is recognized for its affordability and operational simplicity. The chemicals involved, such as aluminum sulfate (alum) and ferric chloride, are inexpensive and widely available. These coagulants efficiently precipitate arsenic without requiring extensive infrastructure or specialized equipment, which minimizes initial setup costs. Furthermore, the ongoing operational expenses associated with coagulation-flocculation are relatively low, mainly involving periodic replenishment of coagulants and routine maintenance of treatment facilities. This cost-effectiveness makes the method particularly attractive for communities and regions where financial resources are limited (Ravenscroft et al. 2009).

Ease of Implementation

Coagulation-flocculation stands out for its ease of implementation across different scales and settings. The process is straightforward and can be integrated into existing water treatment facilities with minimal modifications. It typically involves mixing the coagulant with water, allowing flocculation to occur through gentle stirring or agitation, and then settling the flocs formed in sedimentation basins. This simplicity extends to the training requirements for operators, who can quickly learn to manage and optimize the process. The robust nature of coagulation-flocculation enables it to adapt to variations in raw water quality and operational conditions, ensuring consistent performance in arsenic removal without the need for complex adjustments (Smith et al. 2000).

Coagulation-flocculation emerges as a robust and reliable method for addressing arsenic contamination in drinking water due to its high efficiency, cost-effectiveness, and ease of implementation. Its ability to achieve significant reduction in arsenic even at low concentrations, coupled with its affordability and operational simplicity, makes it a preferred choice for water treatment facilities worldwide. Taking advantage of these strengths, coagulation-flocculation continues to play a key role in safeguarding public health and ensuring access to safe drinking water in arsenic-affected communities.

11.7 Limitations of Coagulation-Flocculation

Chemical Use and Potential Secondary Pollution

Coagulation-flocculation, while effective, raises concerns about chemical usage and potential secondary pollution. The process relies on the addition of coagulants such as aluminum sulfate (aluminum) or ferric chloride to water. These chemicals work by precipitating arsenic and other contaminants into larger flocs, which are then removed by sedimentation or filtration. However, excessive use of coagulants can lead to residual concentrations of aluminum or iron in treated water, exceeding regulatory limits, and potentially impacting aquatic ecosystems downstream (Ravenscroft et al. 2009).

In addition, the chemical reactions involved in coagulation-flocculation can alter water chemistry, affecting pH levels, and introducing new chemical species into the environment. The presence of these chemicals in effluent water requires careful monitoring and management to prevent unintended environmental consequences and ensure compliance with water quality standards (Smedley and Kinniburgh 2002).

Sludge Production and Disposal Challenges

A significant challenge associated with coagulation-flocculation is the generation and disposal of sludge during the treatment process. As coagulants react with water to form flocs, these flocs accumulate as sludge in sedimentation basins or filtration systems. The sludge typically contains concentrated levels of arsenic, heavy metals, and other contaminants removed from the water, making its disposal a critical environmental concern.

Effective management of sludge removal involves several considerations, including dewatering, stabilization, and safe handling to prevent recontamination of water sources or soil. Depending on the composition and toxicity of the sludge, disposal options can include landfilling in specialized facilities or use in beneficial applications such as agricultural soil amendment (Smith et al. 2000).

Although coagulation-flocculation is a widely adopted method for the removal of arsenic due to its effectiveness and operational feasibility, it presents challenges related to chemical use and sludge management. Addressing these limitations requires integrated approaches that prioritize environmental sustainability and public health protection. Future research efforts should focus on optimizing coagulant dosages, developing alternative treatment technologies, and advancing sludge treatment and disposal techniques to improve the overall efficiency and sustainability of water treatment processes.

11.8 Adsorption

Adsorption is a key process in water treatment strategies that aims to remove arsenic and various other contaminants from aqueous solutions. This method operates on the principle of attracting arsenic ions (As(III) As(III) and As(V) As(V)) to the surface of adsorbent materials through physical or chemical interactions. Commonly employed adsorbents include activated carbon, iron oxides (such as iron oxide-coated sand), zeolites, and modified clays, all of which feature extensive surface areas and specific chemical properties conducive to effectively binding arsenic (Chowdhury et al. 2016).

Activated carbon, which is known for its porous structure and large surface area, is particularly effective in adsorbing both forms of arsenic. Iron oxides, either naturally occurring or synthesized, exhibit a strong affinity for arsenic through surface complexation reactions. These adsorbents are used in various configurations such as fixed-bed columns, batch reactors, or filtration systems, offering flexibility in water treatment applications to address arsenic contamination (Mohan and Pittman 2007).

The efficiency of adsorption processes is influenced by several key factors. pH levels play a critical role, affecting the charge of both the adsorbent surface and the arsenic species. Generally, arsenate (As(V) As(V)) adsorption increases under acidic conditions, while arsenite (As(III) As(III)) is more adsorbable under alkaline conditions. The contact time between the adsorbent and water is another crucial parameter; longer durations allow for greater arsenic uptake as more ions interact with the available surface sites.

Additionally, the initial concentration of arsenic in the water source impacts the efficiency of adsorption. Higher concentrations may saturate adsorption sites more quickly, necessitating adjustments in the adsorbent dosage to maintain effective treatment. Temperature also affects adsorption kinetics, typically enhancing reaction rates at higher temperatures, although this effect varies among different adsorbent materials and arsenic species. Furthermore, the presence of competing ions in water, such as sulfates, phosphates, and organic matter, can interfere with arsenic adsorption by competing for binding sites on the adsorbent surface.

Adsorption emerges as a versatile and effective method for mitigating arsenic contamination in water because of its robust ability to selectively capture arsenic ions from aqueous solutions. Understanding the mechanisms and optimizing the operational parameters influencing the adsorption capacity is essential for designing and implementing efficient water treatment systems. Continued research efforts and technological advancements in adsorption processes hold promise for further improving arsenic removal efficiency and ensuring access to safe drinking water in arsenic-affected regions.

11.9 Strengths of Adsorption for Arsenic Removal

Adsorption stands out as a highly effective method of removing arsenic from water because of several distinct strengths, making it a preferred choice in water treatment strategies.

High Removal Efficiency for Both As(III) and As(V)

One of the primary strengths of adsorption is its ability to achieve high removal efficiencies for both arsenite (As(III)) and arsenate (As(V)) species from aqueous solutions. Adsorbent materials, such as activated carbon, iron oxides, and zeolites, possess specific surface properties and chemical structures that facilitate strong interactions with arsenic ions. This enables them to effectively bind and immobilize arsenic contaminants, reducing their concentration in treated water to levels that meet regulatory standards for safe drinking water (Chowdhury et al. 2016).

Simplicity and Ease of Operation

Adsorption processes are characterized by their operational simplicity and ease of implementation on various scales of water treatment facilities. The method typically involves passing contaminated water through a bed or filter containing the adsorbent material. This straightforward process requires minimal infrastructure and can be adapted for use in both centralized treatment plants and decentralized systems in remote or rural areas. In addition, adsorption systems are relatively robust and can operate effectively under a wide range of water quality conditions, making them versatile for treating arsenic-contaminated water from various sources (Mohan and Pittman 2007).

The adsorption offers significant strengths in the removal of arsenic, including high efficiency in the removal of both As(III) and As(V) species and operational simplicity. These attributes underscore its effectiveness as a viable solution to address arsenic contamination in water supplies. As research and technological advancements continue to improve adsorbent materials and optimize treatment processes, adsorption remains a cornerstone for ensuring access to safe and clean drinking water in arsenic-affected regions.

11.10 Limitations of Adsorption for Arsenic Removal

While adsorption is highly effective, it also comes with certain limitations that affect its practical application in the removal of arsenic from water sources.

Need for Periodic Regeneration or Replacement of Adsorbents

One significant limitation of adsorption-based arsenic removal systems is the requirement for periodic regeneration or replacement of adsorbent materials. Over time, adsorbents become saturated with arsenic and other contaminants, reducing their adsorption capacity and effectiveness. Regeneration processes, such as chemical treatment or thermal desorption, may be necessary to restore adsorbent performance. However, these processes can be complex, costly and may involve handling hazardous chemicals, posing challenges for operational maintenance in remote or resource-limited settings (Chowdhury et al. 2016).

High Cost of High-Quality Adsorbents

Another critical limitation is the high cost associated with acquiring and maintaining high-quality adsorbents suitable for arsenic removal. Adsorbent materials, such as activated carbon or specialized iron oxides, are often expensive because of their manufacturing processes and specific properties that enhance arsenic adsorption capacity. The initial investment in procuring these adsorbents, along with ongoing operational costs for regeneration or replacement, can impose financial constraints on the implementation of adsorption-based treatment systems, particularly in economically disadvantaged regions (Mohan and Pittman 2007).

Adsorption is recognized for its effectiveness in arsenic removal; however, it faces limitations related to the periodic need for regeneration or replacement of adsorbents and the high cost of acquiring high-quality materials. Addressing these limitations requires ongoing research and development efforts to optimize adsorbent performance, reduce operational costs, and explore alternative regeneration methods. Despite these challenges, adsorption remains a viable technology for ensuring safe drinking water by mitigating arsenic contamination, although it has considerations for sustainable management and cost-effective implementation.

11.11 Ion Exchange

Mechanism and Operational Principles

Ion exchange is a widely used water treatment method that relies on the principle of exchanging ions in solution with ions bound to a solid, typically resin-based material. In the context of arsenic removal, ion exchange involves the passage of arsenic-contaminated water through a column packed with resin beads. These resin beads are composed of synthetic organic polymers, such as polystyrene, which are functionalized with specific groups known as functional ligands or ion exchange sites.

Description of Ion Exchange Processes and Resin Types

The ion exchange process begins as water flows through the resin bed. Arsenic ions (As(III)\{As(III)}As(III) and As(V)\{As(V)}As(V)) in the water phase are attracted to and adsorb onto the resin beads. Meanwhile, counterions (e.g., chloride ions) that were originally bound to the resin are released into the water in exchange for the arsenic ions. This exchange mechanism occurs because of electrostatic forces and chemical affinity between the functional groups (e.g., quaternary ammonium or quaternary phosphonium groups in strong base anion exchange resins) and the arsenic species.

Various types of ion exchange resins are tailored for specific water treatment applications. Strong base anion exchange resins are commonly used for arsenic removal because they possess positively charged functional groups that attract negatively charged arsenic ions. These resins effectively capture both As(III)\{As(III)}As(III) and As(V)\{As(V)}As(V) species from water, ensuring comprehensive arsenic removal (Mohammad et al. 2020).

Selectivity and Efficiency in Arsenic Removal

Ion-exchange resins exhibit high selectivity for arsenic ions, enabling efficient removal even at low concentrations. The selectivity arises from its specific affinity for arsenic species over competing ions like sulfates, chlorides, and bicarbonates. This selectivity ensures that arsenic is effectively removed from water without significant interference from other ions present in natural water sources.

The efficiency of arsenic removal via ion exchange depends on several factors, including the resin's capacity, regeneration potential, and operational conditions such as flow rate and pH. Regeneration of ion exchange resins involves eluting captured arsenic ions using chemical solutions, allowing the resin to be reused multiple times before replacement is necessary. However, the regeneration process can be resource-intensive and requires careful management to maintain optimal resin performance over time.

Ion exchange represents a robust and selective technology for mitigating arsenic contamination in water supplies. Using the specific binding properties of resin-based materials, ion exchange systems effectively reduce arsenic concentrations to meet regulatory standards for safe drinking water. Continued research and technological advancements in resin chemistry and process optimization are crucial to enhancing ion exchange efficiency, reducing operational costs, and expanding access to clean water in arsenic-affected regions.

11.12 Strengths of Ion Exchange for Arsenic Removal

Ion exchange is a highly effective method for removing arsenic from water, offering several distinct strengths that make it a preferred technology in water treatment.

High Selectivity for Arsenic Ions

One of the primary strengths of ion exchange is its exceptional selectivity for arsenic ions (As(III)\{As(III)}As(III) and As(V)\{As(V)}As(V)). Ion exchange resins are designed with specific functional groups that attract and selectively bind arsenic ions to other competing ions present in water, such as sulfates or chlorides. This selective affinity ensures that arsenic is efficiently removed from water even at low concentrations, providing a reliable method to achieve stringent water quality standards (Mohammad et al. 2020).

Effective at Low Arsenic Concentrations

Ion exchange systems are particularly effective in treating water sources with low to moderate arsenic concentrations. Resin-based materials used in ion exchange processes have a high capacity for arsenic adsorption, allowing them to effectively capture trace amounts of arsenic from large volumes of water. This capability is crucial for ensuring safe drinking water in regions where arsenic contamination levels fluctuate or where strict regulatory limits must be met (Nguyen and Vigneswaran 2009).

The ion exchange demonstrates high selectivity for arsenic ions and is effective at treating low concentrations of arsenic in water. These strengths underscore the role of ion exchange as a robust and reliable technology to mitigate arsenic contamination and ensure safe drinking water supplies. Ongoing advances in resin chemistry and process optimization continue to enhance the efficiency and affordability of ion exchange systems, further expanding their application in arsenic-affected regions.

11.13 Limitations of Ion Exchange for Arsenic Removal

While effective, is subject to certain limitations that affect its practical application in the removal of arsenic removal from water sources.

Requirement for Resin Regeneration

One significant limitation of ion exchange systems is the need for periodic resin regeneration. Over time, the resin beads used in ion exchange become saturated with arsenic and other ions removed from water. Regeneration involves treating the resin with a chemical solution to release and flush out captured ions, including arsenic, thereby restoring the resin's capacity for further use. However, this regeneration process can be resource intensive, requiring chemicals and water to rinse, and may pose challenges in terms of operational downtime and maintenance logistics (Nguyen and Vigneswaran 2009).

Impact of Competing Ions on Effectiveness

The effectiveness of ion exchange for arsenic removal can be influenced by the presence of competing ions in the water supply. Competing ions, such as sulfates, chlorides, and bicarbonates, can compete with arsenic ions for binding sites on the resin surface. This competition may reduce the selectivity for arsenic and lead to decreased removal efficiency, especially in water sources with complex chemical compositions or high concentrations of competing ions. Managing and mitigating the impact of competing ions through proper resin selection and operational adjustments is critical for maintaining optimal arsenic removal performance (Mohammad et al. 2020).

The ion exchange offers high selectivity and effectiveness in arsenic removal from water, but it faces limitations related to the periodic need for resin regeneration and the impact of competing ions on treatment efficiency. Addressing these limitations requires ongoing research and development efforts to optimize resin regeneration processes, enhance resin selectivity for arsenic ions, and improve overall system performance under varying water quality conditions. Despite these challenges, ion exchange remains a valuable technology for ensuring safe drinking water by effectively reducing arsenic contamination levels.

11.14 Mechanisms and Operational Principles of Membrane Filtration

Membrane filtration is a versatile separation process that employs semipermeable membranes to selectively separate substances on size and molecular weight. The mechanism involves passing a feed solution through the membrane, where particles larger than the membrane's pore size are retained while smaller molecules pass through. This process can be driven by pressure differentials, such as in pressure-driven processes such as microfiltration (MF), ultrafiltration (UF), and reverse osmosis (RO), or by concentration gradients, such as in osmotically driven processes like nanofiltration (NF) and membrane distillation (MD).

Description of Membrane Filtration Techniques

Microfiltration (MF) uses membranes with pore sizes typically range from 0.1 to 10 μm. These membranes are effective at removing suspended solids, bacteria, and some colloidal materials from liquids. Applications of MF include clarification of beverages, pretreatment for subsequent filtration processes, and wastewater treatment to achieve particle removal and reduction in turbidity.

Ultrafiltration (UF) membranes have smaller pore sizes, typically between 0.001 and 0.1 μm, allowing the removal of dissolved macromolecules, proteins, colloids, and pyrogens. UF finds extensive use in the dairy industry for the concentration and fractionation of milk components, as well as in pharmaceutical manufacturing for protein purification and elimination of endotoxins from solutions.

Nanofiltration membranes (NF) feature even smaller pores, ranging from 0.001 to 0.01 μm, and exhibit selective permeability to divalent ions, organic compounds, and smaller salts. NF is particularly valuable in water treatment processes for softening hard water, removal of color from effluents, and selective removal of specific contaminants while allowing the passage of water and smaller ions.

Reverse osmosis (RO) membranes have the smallest pore sizes, typically less than 0.001 μm, and are capable of effectively removing salts, ions, and organic molecules from water streams. RO technology is indispensable for seawater desalination and brackish water, production of ultrapure water for industrial processes, and the concentration of juices and dairy products.

Membrane distillation (MD) involves hydrophobic membranes that facilitate the selective transport of water vapor through vapor pressure differences, separating pure water from saline solutions or wastewater. MD is advantageous for desalination in remote areas, concentration of heat-sensitive products, and treatment of industrial effluents where traditional methods may be less effective.

Membrane filtration technologies encompass a range of techniques and membrane types tailored to specific separation needs in various industries. Understanding the operational principles and capabilities of microfiltration, ultrafiltration, nanofiltration, reverse osmosis, and membrane distillation is crucial for optimizing processes in water treatment, food and beverage production, pharmaceutical manufacturing, and environmental remediation. As advances in membrane materials and manufacturing processes continue, these technologies promise to deliver more efficient, sustainable, and cost-effective solutions for complex separation challenges in the future.

11.15 Strengths of Membrane Filtration

High Removal Efficiencies

Membrane filtration processes are renowned for their exceptional removal efficiencies across a spectrum of contaminants. Microfiltration membranes (MF), with pore sizes typically ranging from 0.1 to 10 μm, effectively remove suspended solids, bacteria, and protozoa from water and wastewater streams. This makes MF ideal for applications such as drinking water treatment, where the removal of pathogens and turbidity is critical to ensure water safety.

Ultrafiltration membranes (UF) operate with smaller pore sizes (0.001–0.1 μm), allowing them to remove macromolecules, colloids, and some viruses from water sources. UF is widely used in industries requiring precise separation and concentration processes, such as dairy processing for protein fractionation and pharmaceutical manufacturing for the purification of biologics.

Nanofiltration (NF) membranes offer selective removal of divalent ions, organic compounds, and smaller salts due to their tighter pore structure (0.001–0.01 μm). NF is particularly effective in water softening applications, where it can selectively remove hardness ions like calcium and magnesium while allowing the passage of water molecules, resulting in softened water suitable for industrial processes and domestic use.

Reverse osmosis (RO) membranes have the smallest pores (<0.001 μm) and can effectively remove dissolved salts, ions, and organic molecules from water through a pressure-driven process. RO technology is critical for the desalination of seawater and brackish water, providing a sustainable source of fresh water in arid regions and coastal areas facing water scarcity.

Versatility in Handling Various Water Qualities

One of the key advantages of membrane filtration is its versatility in treating diverse water qualities and varying feed conditions. MF and UF membranes are robust against fluctuations in feed water quality, capable of handling turbid or particle-laden waters without compromising filtration performance. This resilience makes them suitable for treating surface water, groundwater, and wastewater from various industrial processes, ensuring consistent water quality output.

NF and RO membranes excel in treating high salinity water sources, removing salts, and achieving high purity water suitable for drinking, industrial processes, and specialized applications such as pharmaceutical manufacturing. Their ability to selectively separate ions and molecules based on size and charge enables precise control over water composition, meeting stringent regulatory standards and operational requirements.

Membrane filtration technologies offer unparalleled strengths in terms of high removal efficiencies, versatility in different water qualities, and suitability for a wide range of applications. These technologies play a key role in ensuring access to safe drinking water, treating industrial effluents, and addressing environmental challenges associated with water pollution and scarcity. As advances continue to enhance membrane materials, system designs, and operational efficiency, membrane filtration remains a cornerstone of sustainable water management strategies worldwide.

11.16 Limitations of Membrane Filtration

High Operating and Maintenance Costs

Despite their effectiveness, membrane filtration systems are associated with high operational and maintenance costs. Initial capital investments to purchase and install membrane systems can be substantial, especially for large-scale applications, such as municipal water treatment plants or industrial facilities. These costs include the procurement of membranes, pumps, and monitoring equipment, as well as expenses related to the design, installation, and commissioning of the system.

Operational costs include energy consumption for maintaining pressure differentials between membranes, as well as ongoing expenses for water treatment chemicals, membrane cleaning agents, and periodic membrane replacement. Routine maintenance activities, including membrane cleaning, integrity testing, and system monitoring, are essential to ensure optimal performance and long-term performance of membrane systems. These costs can pose financial challenges, particularly for smaller municipalities or industries operating under tight budget constraints.

Membrane Fouling and Its Impact on Efficiency

Membrane fouling remains a significant operational challenge in membrane filtration processes, affecting system efficiency and performance over time. Fouling occurs when contaminants, organic matter, scaling agents, or microorganisms accumulate on the membrane surface or within its pores, leading to reduced flux rates, increased energy consumption, and reduced water quality output.

Various factors contribute to membrane fouling, including feed water composition, operating conditions, and membrane material properties. Particulate fouling occurs as a result of the deposition of suspended solids on the membrane surface, while organic fouling results from the adsorption of organic molecules and biofilm formation. Scaling fouling arises from the precipitation of salts and minerals on the membrane surface, particularly in RO and NF processes dealing with high-salinity waters.

Mitigating membrane fouling requires proactive measures such as regular cleaning protocols using appropriate chemicals, optimizing operating conditions to minimize fouling propensity, and implementing pretreatment processes like coagulation, flocculation, and microfiltration to reduce feed water contaminants. Research efforts continue to focus on developing antifouling membranes, advanced cleaning techniques, and predictive fouling models to improve operational efficiency and extend membrane lifespan.

The membrane filtration technologies offer substantial benefits in terms of water quality improvement and resource conservation, but they are accompanied by inherent limitations such as high operational and maintenance costs, as well as challenges related to membrane fouling. Addressing these limitations requires ongoing technological innovations, operational optimizations, and cost-effective management strategies to maximize the sustainability and viability of membrane filtration systems in diverse applications. Future advancements in membrane materials, fouling mitigation techniques, and system design are essential to further improve the efficiency and affordability of membrane-based water treatment solutions.

11.17 Oxidation: Mechanism and Operational Principles

Oxidation is a chemical process that involves the transfer of electrons from a substance to an oxidizing agent, resulting in an increase of the oxidation state of the substance. In water treatment, oxidation processes are crucial in transforming contaminants into more easily removable forms, enhancing the efficiency of subsequent treatment steps.

Description of Oxidation Processes and Agents Used

Various oxidation processes and agents are employed to convert contaminants like arsenic (As) from less soluble or more toxic forms to more readily removable forms. One of the significant applications is the conversion of arsenite (As(III)), which is more toxic and less easily removable, to arsenate (As(V)), a less toxic and more easily removable species.

Chemical Oxidation: Chemical oxidation involves the use of oxidizing agents such as chlorine (Cl_2), ozone (O_3), potassium permanganate ($KMnO_4$), and hydrogen peroxide (H_2O_2). These agents react with As(III) in water, facilitating the oxidation reaction:

$$\text{As(III)} + \text{Oxidizing agent} \rightarrow \text{As(V)} + \text{Reduced product}$$

Chlorine is effective in oxidizing As(III) to As(V) under controlled pH conditions, making it suitable for large-scale water treatment applications (Sorg et al. 2014). Ozone, known for its strong oxidative properties, rapidly oxidizes As(III) without producing harmful by-products, making it advantageous in drinking water treatment where minimizing chemical residues is critical (Guo et al. 2011).

Biological Oxidation: Biological oxidation processes utilize microorganisms to catalyze oxidation reactions. For arsenic, biological oxidation primarily involves the microbial-mediated oxidation of As(III) to As(V). Certain bacteria, such as species of

Thiobacillus, can oxidize As(III) as part of their metabolic processes, converting it to less toxic forms that can be subsequently removed through filtration or precipitation processes (Mandal and Suzuki 2002).

Conversion of As(III) to More Easily Removable As(V)

The oxidation of As(III) to As(V) is a critical step in arsenic removal processes, as As(V) forms complexes with iron and other metals, facilitating its removal by coagulation, precipitation, or filtration. As(V) is also less soluble in water compared to As(III), making it easier to separate from the aqueous phase through physical or chemical treatment processes (Smedley and Kinniburgh 2002).

Oxidation processes play a vital role in water treatment by converting harmful contaminants such as As(III) into less toxic and more easily removable forms such as As(V). Understanding the mechanisms and operational principles of oxidation, along with the selection of appropriate oxidation agents, is crucial for designing effective treatment strategies to ensure safe drinking water and environmental protection. Future research and technological advances will continue to improve oxidation processes, improving their efficiency, sustainability, and applicability to address emerging challenges in water quality challenges globally.

11.18 Strengths of Oxidation Processes in Water Treatment

Effective Oxidation of Arsenic Species

Oxidation processes are highly effective in the transformation of arsenic species, particularly in the conversion of arsenite (As(III)), which is more toxic and less easily removable, into arsenate (As(V)), a less toxic and more readily removable form. This transformation improves the efficiency of subsequent treatment steps, such as coagulation, filtration, or precipitation, which target the removal of As(V) from water sources (Smedley and Kinniburgh 2002).

Chemical oxidants, such as chlorine, ozone, potassium permanganate, and hydrogen peroxide, are commonly used for arsenic oxidation because of their ability to react with As(III) and facilitate the oxidation reaction. These agents provide reliable and rapid oxidation under controlled conditions, ensuring consistent treatment performance with varying water qualities and operational parameters (Guo et al. 2011; Sorg et al. 2014).

Integration with Other Treatment Methods

Another significant strength of oxidation processes is their compatibility and integration with various other water treatment methods. After As(III) to As(V), the treated water can undergo further purification steps, such as filtration through activated carbon, ion exchange, or membrane processes. This integration allows for the comprehensive removal of arsenic and other contaminants, tailoring treatment strategies to meet specific water quality standards and regulatory requirements (Mandal and Suzuki 2002).

For instance, oxidation followed by coagulation and filtration processes effectively removes As(V) by forming precipitates with iron or aluminum-based coagulants, which are then removed through settling or filtration. This multi-barrier approach enhances treatment efficiency and ensures the production of safe drinking water or compliant effluent for discharge into the environment (Smedley and Kinniburgh 2002).

The strengths of oxidation processes, including effective oxidation of arsenic species and their integration with complementary treatment methods, highlight their critical role in arsenic removal and water treatment applications. By leveraging the capabilities of chemical and biological oxidants, water treatment facilities can achieve robust and reliable performance in addressing arsenic contamination challenges, ensuring the delivery of safe and clean water resources to communities around the world. Continued advancements in oxidation technologies and treatment strategies will further enhance their effectiveness and sustainability in mitigating emerging water quality issues.

11.19 Limitations of Oxidation Processes in Water Treatment

Chemical Use and Associated Costs

One of the primary limitations of oxidation processes is the reliance on chemical oxidants such as chlorine, ozone, potassium permanganate, and hydrogen peroxide. These chemicals are essential for the effective oxidation of contaminants such as arsenic, but contribute to operational costs due to their procurement, storage, and dosing requirements. Initial capital investment and ongoing operational expenses associated with chemical usage can be significant, particularly for large-scale water treatment facilities or areas with high arsenic concentrations (Guo et al. 2011).

In addition, the cost-effectiveness of oxidation processes depends on factors such as the concentration of contaminants in the feed water, the required treatment levels, and regulatory compliance standards. Balancing treatment efficacy with economic feasibility poses challenges, especially in regions with limited financial resources or competing priorities for infrastructure development (Sorg et al. 2014).

Potential Environmental Impacts

The use of chemical oxidants in water treatment processes may pose potential environmental impacts, primarily related to the discharge of residual chemicals and their by-products into water bodies. Chlorine, for example, can form disinfection by-products (DBPs) such as trihalomethanes (THMs) and haloacetic acids (HAAs) when reacting with organic matter in water. These DBPs are regulated due to their potential health risks and environmental persistence, necessitating careful management and monitoring to mitigate adverse impacts (Richardson et al. 2007).

Although effective as an oxidant, requires energy-intensive generation processes, leading to increased carbon footprints and operational costs. The discharge of ozone residuals into aquatic environments can also impact aquatic ecosystems and aquatic life, requiring comprehensive environmental risk assessments and mitigation measures (Guo et al. 2011).

The oxidation processes offer effective solutions for transforming contaminants like arsenic into more manageable forms, and they are accompanied by limitations related to chemical use and associated costs, as well as potential environmental impacts. Addressing these challenges requires a balanced approach that considers technological advances in chemical-free alternatives, improved process efficiencies, and sustainable management practices to ensure safe and environmentally responsible water treatment solutions. Continued research and innovation will play a crucial role in advancing oxidation technologies while minimizing their environmental footprint and enhancing overall water quality management strategies.

11.20 Comparative Analysis of Oxidation Methods in Water Treatment

Chemical Oxidation

Strengths: Chemical oxidation involves the use of oxidizing agents such as chlorine, ozone, potassium permanganate, and hydrogen peroxide. These agents are highly effective in rapidly oxidizing contaminants such as arsenic (As(III)) to less toxic forms such as arsenate (As(V)), which are easier to remove through subsequent treatment processes such as filtration or precipitation (Guo et al. 2011; Sorg et al. 2014). Chemical oxidation methods are versatile and can be applied in a wide range of water qualities and contaminant concentrations, making them suitable for both municipal and industrial water treatment applications.

Limitations: One significant limitation of chemical oxidation methods is the reliance on chemical agents, which can contribute to high operational costs as a result of procurement, storage, and dosing requirements. The formation of disinfection by-products (DBPs) during chlorine and ozone oxidation processes also poses environmental concerns, requiring careful management to minimize impacts on water quality and aquatic ecosystems (Richardson et al. 2007). Additionally, the effectiveness of chemical oxidation can vary under different water quality conditions and pH levels, which affects treatment performance and requires adjustments in chemical dosage and treatment protocols.

Biological Oxidation

Strengths: Biological oxidation processes utilize microorganisms to catalyze oxidation reactions, offering a natural and sustainable approach to the removal in water treatment. These processes can effectively degrade organic pollutants and transform some metal ions under suitable environmental conditions, reducing the reliance on chemical inputs and minimizing environmental impacts (Mandal and Suzuki 2002). Biological oxidation is generally cost-effective once established because it relies on microbial activity for oxidation without continuous chemical additions.

Limitations: However, biological oxidation methods are sensitive to environmental factors such as temperature, pH, and nutrient availability, which can affect microbial activity and treatment efficiency. This sensitivity may require longer residence times or more controlled conditions to achieve desired treatment outcomes compared to chemical oxidation methods. Biological processes also have limitations in treating certain contaminants or in environments with fluctuating water quality parameters, limiting their application in specific water treatment scenarios (Mandal and Suzuki 2002).

Comparative Effectiveness Under Different Conditions

Contaminant Type and Concentration: Chemical oxidation methods excel in treating a wide range of contaminants, including metals, organic compounds, and pathogens, at varying concentrations. They offer rapid and reliable treatment capabilities, which make them suitable for emergency responses or situations requiring immediate removal of contaminants. Biological oxidation is particularly effective in treating organic pollutants and certain metal ions under stable environmental conditions, where microbial populations can thrive and maintain treatment efficiency over time.

Water Quality and Source: Chemical oxidation methods provide operational flexibility in adjusting treatment parameters to meet the demands of fluctuating water quality and regulatory standards. They are preferred in scenarios requiring precise control over treatment outcomes and compliance with stringent water quality criteria. Biological oxidation processes are advantageous in eco-sensitive areas or applications where sustainable water management practices are prioritized, minimizing chemical use, and reducing environmental footprints.

Operational Considerations: Chemical oxidation methods may require higher initial capital investment and ongoing operational costs due to chemical procurement and disposal requirements. In contrast, biological oxidation processes offer potential cost savings once established, although they may require longer treatment times or additional infrastructure for optimal performance.

The choice between chemical and biological oxidation methods depends on specific operational needs, contaminant characteristics, and environmental considerations in water treatment. Chemical oxidation offers rapid and versatile contaminant removal capabilities but requires careful management of chemical usage and potential environmental impacts. Biological oxidation provides sustainable treatment solutions with lower operating costs but requires favorable environmental conditions and may be limited in the treatment of certain contaminants or under fluctuating water quality conditions. Integrated approaches that combine both oxidation methods or synergize with complementary treatment technologies can optimize treatment efficiency and address diverse water quality effectively, ensuring the provision of safe and reliable water supplies for communities around the world. Continued research and technological advancements will further enhance the comparative effectiveness and sustainability of oxidation methods in mitigating emerging water quality issues globally.

Table 11.1 focuses on each arsenic removal method and its respective advantages and limitations.

Coagulation-Flocculation: This method is highly effective at removing arsenic from water with high concentrations, because of its ability to aggregate arsenic particles into larger clumps (floccules) that can be easily separated from the water. It is widely applicable and relatively easy to implement in water treatment plants. However, coagulation-flocculation requires precise control of pH and turbidity levels to ensure optimal performance. Variations in these parameters can reduce treatment efficiency, highlighting the sensitivity to water quality conditions. Furthermore, the process generates sludge as a byproduct, which requires proper disposal methods. The management of sludge can increase operational costs and pose environmental challenges, which makes it necessary to consider the overall sustainability of the treatment process.

Adsorption (e.g. iron oxide): Adsorption methods involve the use of materials like iron oxide to bind arsenic ions from water onto their surfaces. This process is highly effective at removing arsenic, even at low concentrations, and is reversible, allowing the regeneration of adsorbent materials. The reversibility of the adsorption processes makes them cost-effective over time, despite the initial investment in adsorbent materials. However, the high operational costs associated with periodic regeneration of

Table 11.1 Comparison of conventional arsenic removal methods

Method	Advantages	Limitations
Coagulation-flocculation	Effective at high arsenic concentrations	pH and turbidity sensitivity, sludge disposal
Adsorption (e.g., iron oxide)	High removal efficiency, reversible	High operational costs, regeneration required
Ion exchange	Selective for arsenate, suitable for low concentrations	High cost of media replacement, brine disposal
Membrane filtration	Physical barrier, effective at low concentrations	Membrane fouling, high operational and energy costs

adsorbents and the need for proper disposal of spent materials are significant considerations. Adsorption methods are particularly suitable for applications where consistent and reliable removal of arsenic is critical, balancing the initial financial outlay with long-term operational benefits.

Ion Exchange: Ion exchange involves the exchange of arsenic ions that are in water with other ions held on a resin material. This method is highly selective for arsenate and is effective in treating water with low to moderate concentrations of arsenic. It can also treat multiple contaminants concurrently, enhancing its utility in various water treatment scenarios. However, ion exchange systems require regular replacement of exhausted resin and careful management of brine waste generated during regeneration cycles. These operational requirements contribute to higher ongoing costs compared to other methods. Additionally, the effectiveness of ion exchange can be influenced by the chemical composition, necessitating pretreatment to optimize performance and extend resin life.

Membrane Filtration: Membrane filtration uses physical barriers to separate arsenic and other contaminants from water. It is particularly effective in treating low concentrations of arsenic without the need for chemical additives, making it environmentally preferable in certain contexts. Membrane filtration systems operate by passing water through membranes with specific pore sizes that selectively trap contaminants, including arsenic. However, membrane fouling, caused by the accumulation of particles and contaminants on the membrane surface, remains a significant operational challenge. Regular maintenance, such as cleaning or replacing membranes, is essential to maintain filtration efficiency over time. Furthermore, membrane filtration systems typically operate at high pressures, resulting in elevated energy consumption and operational costs.

Each method presented in Table 11.1 offers unique strengths and challenges in the context of the removal of arsenic from water. Coagulation-flocculation and adsorption methods are highly effective but require careful management of operational conditions and costs, including sludge disposal and regeneration processes. Ion exchange provides selective removal but entails ongoing investments in resin replacement and brine waste management. Membrane filtration offers reliable treatment at low concentrations of arsenic but faces challenges related to fouling and energy consumption. Choosing the appropriate method depends on balancing these factors with specific water quality requirements and operational capabilities, ensuring effective and sustainable arsenic remediation practices.

11.21 Cost–Benefit Analysis of Oxidation Methods in Water Treatment

Operational and Maintenance Costs

Chemical Oxidation: Chemical oxidation methods involve the use of oxidizing agents such as chlorine, ozone, potassium permanganate, and hydrogen peroxide. Although these methods offer rapid and effective contaminant removal, they incur significant operational and maintenance costs. The procurement, storage, and dosing of chemicals contribute to ongoing expenses, particularly in large-scale water treatment facilities or areas with high concentrations (Guo et al. 2011). Additionally, the management of chemical residuals and disinfection by-products (DBPs) requires continuous monitoring and treatment, adding to operational complexities and costs (Richardson et al. 2007).

Biological Oxidation: In contrast, biological oxidation processes utilize microbial activity to catalyze oxidation reactions, offering potential cost savings in chemical input once established. Operational costs are generally lower than those of chemical oxidation methods, as they rely on natural biological processes for contaminant removal. However, biological oxidation may require initial investments in infrastructure and monitoring systems to optimize microbial activity and treatment performance (Mandal and Suzuki 2002). Long-term operational costs include maintenance of biological reactors and periodic monitoring to ensure stable treatment efficiency.

Assessment of Long-Term Sustainability and Feasibility

Chemical Oxidation: Chemical oxidation methods provide immediate and reliable contaminant removal capabilities, making them suitable for applications requiring rapid treatment responses or compliance with stringent water quality standards. Despite higher initial costs and ongoing chemical expenditures, these methods are well-established and widely used in municipal and industrial water treatment facilities. The feasibility of chemical oxidation depends on factors such as the availability of skilled personnel, regulatory compliance requirements, and the ability to manage chemical residuals and DBPs effectively (Sorg et al. 2014).

Biological Oxidation: Biological oxidation offers sustainable treatment solutions with lower chemical use and reduced environmental impacts compared to chemical oxidation methods. Natural degradation processes facilitated by microorganisms improve long-term treatment sustainability and compatibility of the ecosystem. However, the feasibility of biological oxidation depends on maintaining optimal environmental conditions for microbial activity, including stable pH levels, appropriate nutrient availability, and consistent temperature ranges (Mandal and Suzuki 2002). Challenges such as sensitivity to environmental fluctuations and longer treatment times may require additional investments in process control and monitoring systems to ensure reliable and consistent treatment performance.

Cost–benefit analysis of oxidation methods in water treatment underscores the importance of considering operational costs, long-term sustainability, and feasibility in selecting appropriate treatment strategies. Chemical oxidation methods offer rapid contaminant removal but entail higher operational and maintenance costs associated with chemical procurement and management of by-products. Biological oxidation provides cost-effective and environmentally friendly treatment alternatives but requires careful management of microbial processes and environmental conditions to optimize treatment efficiencies. Integrated approaches that combine both oxidation methods or integrate biological processes with advanced treatment technologies can enhance treatment performance and address diverse water quality challenges effectively, ensuring sustainable and resilient water supply systems for communities around the world. Continued advancements in oxidation technologies and treatment strategies will further optimize cost-effectiveness and environmental stewardship in water treatment practices.

11.22 Future Directions in Arsenic Removal Technologies

Research Gaps

Identification of Areas Needing Further Investigation: There remains a critical need for comprehensive studies to address several research gaps in arsenic removal technologies. First, understanding the long-term impacts of existing treatment methods on water quality and public health is essential. Research should focus on assessing the effectiveness of current technologies in different contexts of water quality and under varying environmental conditions. Furthermore, it is a need to explore the mechanisms underlying arsenic mobilization, transformation, and transport in groundwater and surface water systems. This includes investigating the role of geochemical processes, microbial interactions, and hydrogeological factors in the dynamics of arsenic contamination.

Potential for Innovation in Arsenic Removal Technologies: The field of arsenic removal technologies has significant potential for innovation. Research efforts should prioritize the development of cost-effective, scalable, and environmentally sustainable solutions. Innovative technologies such as nanoscale materials, advanced oxidation processes, and hybrid treatment systems show promise in enhancing arsenic removal efficiencies and reducing treatment costs. Furthermore, exploring the potential of natural sorbents, biofiltration methods, and membrane technologies could lead to breakthroughs in the removal of arsenic from water sources.

Innovative Solutions

Emerging Technologies and Their Potential Impact: Emerging technologies in arsenic removal, such as electrochemical methods, adsorption onto novel materials, and membrane based separation techniques, offer exciting opportunities for improving treatment efficiency and reducing operational footprints. For example, advanced oxidation processes (AOPs) like UV/H_2O_2 and ozone-based treatments are being explored for their ability to oxidize arsenic species effectively and mitigate by-product formation. These technologies hold promise for decentralized water treatment applications and community-level arsenic remediation initiatives.

Integration of Conventional Methods with Advanced Techniques: Integrating conventional arsenic removal methods with advanced techniques represents a promising approach to achieving synergistic treatment outcomes. Hybrid systems that combine adsorption media with membrane filtration or ion exchange resins with AOPs can enhance arsenic removal efficiencies while addressing challenges such as fouling and operational variability. Furthermore, the use of digital technologies, real-time monitoring systems, and data analytics can optimize process control, improve treatment reliability, and facilitate adaptive management strategies in arsenic-affected regions.

Advancement of arsenic removal technologies requires concerted research efforts to address existing gaps and capitalize on innovative solutions. By identifying research priorities, fostering interdisciplinary collaborations, and embracing emerging technologies, stakeholders can accelerate progress towards sustainable arsenic mitigation strategies. Continued investment in research and development, coupled with proactive policy measures and community involvement, will be instrumental in protecting water quality and public health from arsenic contamination around the world.

Table 11.2 presents an overview of emerging technologies for arsenic removal, highlighting their key features and potential applications in water treatment.

Advanced oxidation processes (AOPs) utilize oxidation reactions to break down and destroy arsenic species in water, making them particularly effective for remediating contaminated groundwater where conventional methods may be insufficient. AOPs offer a robust solution by accelerating chemical reactions that transform arsenic into less harmful forms, thus improving water quality.

Nanoscale materials are characterized by their enhanced surface area, which allows for efficient adsorption of arsenic from water. These materials are ideal for point-of-use water treatment systems, where their compact size and high adsorption capacity provide effective and reliable arsenic removal. The use of nanoscale materials represents a technological advancement in water treatment, offering a practical

Table 11.2 Emerging technologies for arsenic removal

Technology	Key features	Potential applications
Advanced Oxidation processes (AOPs)	Destroys arsenic species through oxidation reactions	Contaminated groundwater remediation
Nano-scale materials	Enhanced surface area for adsorption	Point-of-use water treatment systems
Biological methods	Microbial-mediated transformation or removal	Sustainable and eco-friendly treatment options
Hybrid systems	Integration of multiple treatment processes	Complex arsenic contamination scenarios

solution for households and communities dealing with arsenic-contaminated water sources.

Biological methods leverage microbial activities to transform or remove arsenic from water. This approach is sustainable and environmentally friendly, aligning with the growing global efforts towards environmentally responsible water treatment practices. Biological methods offer versatility under various environmental conditions, providing a viable alternative or complement to chemically intensive arsenic removal technologies.

Hybrid systems integrate multiple arsenic removal techniques into a single system, combining the strengths of different methods to address complex contamination scenarios effectively. By synergizing complementary treatment processes, hybrid systems improve overall arsenic removal efficiency and adaptability. This makes them suitable for managing various challenging water quality issues, including arsenic contamination in both urban and rural settings.

Table 11.2 underscores the diversity and innovation in arsenic removal technologies, each catering to specific water treatment needs and environmental considerations. Advanced oxidation processes excel in remediating groundwater, nanoscale materials optimize point-of-use systems, biological methods offer sustainable solutions, and hybrid systems provide comprehensive approaches for complex contamination challenges. Choosing the appropriate technology depends on factors such as water quality parameters, infrastructure capabilities, and the desired level of environmental impact, ensuring effective arsenic mitigation strategies tailored to local conditions.

11.23 Conclusion

Throughout this review, we have dived into the diverse landscape of arsenic removal methods from water sources, synthesizing key insights and findings from existing literature and studies. Conventional methods such as coagulation-flocculation, adsorption onto iron-based media, and ion exchange have emerged as stalwart approaches in mitigating arsenic contamination. These methods have demonstrated significant efficacy in reducing arsenic levels below regulatory thresholds, particularly in controlled laboratory environments and under optimized conditions.

However, the translation of laboratory success to real-world applications presents formidable challenges. Issues such as variability in water chemistry, operational complexities, maintenance requirements, and treatment residual management pose substantial hurdles to the widespread adoption and sustained effectiveness of conventional arsenic removal technologies. The effectiveness of these methods is often contingent on factors like pH, competing ions, and seasonal variations, necessitating adaptive strategies for reliable performance across diverse geographical and hydrological contexts.

Conventional arsenic removal methods represent fundamental pillars in arsenic mitigation efforts, providing critical solutions to improve water quality and safeguarding public health. Their deployment has significantly contributed to the reduction of arsenic-related health risks in affected communities around the world. Despite their efficacy, continuous optimization and innovation are imperative to enhance their reliability, scalability, and cost effectiveness in both centralized and decentralized water treatment systems.

11.24 Recommendations

Practical Applications and Policy Implications: Practical applications should prioritize the development and deployment of decentralized arsenic treatment solutions tailored to local water quality profiles and socioeconomic contexts. Strengthening institutional capacities, promoting community participation, and fostering partnerships between stakeholders are essential to ensure sustainable arsenic mitigation strategies. Policy frameworks must emphasize stringent regulatory standards, technology transfer mechanisms, and incentives for the adoption of innovative arsenic removal technologies.

Directions for Future Research and Development: Future research efforts should focus on the advancement of next-generation arsenic removal technologies that integrate robustness, efficiency, and environmental sustainability. Promising avenues include the exploration of nanoscale materials, advanced oxidation processes, membrane-based technologies, and hybrid treatment systems. Interdisciplinary collaborations, coupled with advances in digital monitoring and predictive modeling, will facilitate the development of adaptive arsenic treatment solutions capable of addressing evolving challenges in water quality.

Long-term studies are warranted to evaluate the life-cycle environmental impacts, economic feasibility, and social acceptance of emerging arsenic removal technologies. Addressing knowledge gaps related to the fate and transport of arsenic species in aquatic environments, as well as the optimization of treatment processes under dynamic operational conditions, will be crucial in guiding evidence-based decision making and fostering resilient water management practices worldwide.

In conclusion, although conventional arsenic removal methods have significantly advanced arsenic mitigation efforts, sustained progress hinges on continuous innovation, collaboration, and adaptive management strategies. By prioritizing research and development, fostering technological innovation and strengthening regulatory frameworks, we can collectively enhance the resilience of water supply systems and ensure equitable access to safe and sustainable drinking water for all populations.

Acknowledgements I extend my sincere gratitude to the Department of Geography, Osmania University, for providing the necessary academic environment, resources, and institutional support that enabled me to conduct this research effectively. The encouragement and facilities offered by the department played a crucial role in shaping the direction and depth of this study. I would also like to express my heartfelt thanks to Dr. Akhilesh Kumar Yadav, Ph.D., MIE, and the Editor(s) of the book, whose expert guidance and critical evaluation have been instrumental in enhancing the academic rigor and quality of this paper. Their meticulous review and valuable suggestions have helped strengthen the arguments and clarity of presentation.

References

Berg, M., Stengel, C., Trang, P. T. K., Viet, P. H., Sampson, M. L., Leng, M., & Fredericks, D. (2001). Magnitude of arsenic pollution in the Mekong and Red River Deltas—Cambodia and Vietnam. *Science of the Total Environment, 273*(1–3), 379–393.

Chen, Y., Graziano, J. H., Parvez, F., Hussain, I., Momotaj, H., van Geen, A., & Ahsan, H. (2011). Arsenic exposure from drinking water and cardiovascular disease in Bangladesh: a prospective cohort study. *British Medical Journal, 342*, d2431.

Chowdhury, S. R., Yanful, E. K., & Pratt, A. R. (2016). Arsenic removal from aqueous solutions by adsorption on magnetite nanoparticles. *Journal of Environmental Management, 166*, 531–539.

Chowdhury, U. K., Biswas, B. K., Chowdhury, T. R., Samanta, G., Mandal, B. K., Basu, G. C., & Chakraborti, D. (2000). Groundwater arsenic contamination in Bangladesh and West Bengal, India. *Environmental Health Perspectives, 108*(5), 393–397.

Guha Mazumder, D. N. (2008). Chronic arsenic toxicity & human health. *Indian Journal of Medical Research, 128*(4), 436–447.

Guo, H., Wang, J., & Hu, X. (2011). Arsenic removal from aqueous solutions by adsorption on iron oxide minerals. *Journal of Environmental Sciences, 23*(8), 1304–1312.

IARC. (2004). Arsenic in Drinking-water. *IARC Monographs on the Evaluation of Carcinogenic Risks to Humans, 84*, 1–477.

Mandal, B. K., & Suzuki, K. T. (2002). Arsenic round the world: A review. *Talanta, 58*(1), 201–235.

Mohammad, A. W., Teow, Y. H., Ang, W. L., Chung, Y. T., & Oatley-Radcliffe, D. L. (2020). Arsenic Removal from Water Using Ion Exchange Resins. In P. Lens, L. Hulshoff Pol, T. K. Sen, & A. P. van Wezel (Eds.), *Environmental Technologies to Treat Nitrogen Pollution* (pp. 311–336). Springer, Cham.

Mohan, D., & Pittman Jr, C. U. (2007). Arsenic removal from water/wastewater using adsorbents—A critical review. *Journal of Hazardous Materials, 142*(1–2), 1–53.

Naujokas, M. F., Anderson, B., Ahsan, H., Aposhian, H. V., Graziano, J. H., Thompson, C., & Suk, W. A. (2013). The broad scope of health effects from chronic arsenic exposure: update on a worldwide public health problem. *Environmental Health Perspectives, 121*(3), 295–302.

Nguyen, T. V., & Vigneswaran, S. (2009). Arsenic Removal by Ion Exchange Processes. In S. Vigneswaran & P. K. T. Venugopalan (Eds.), *Emerging Membrane Technology for Sustainable Water Treatment* (pp. 251–284). IWA Publishing.

Paul, D. S., Walton, F. S., & Thomas, D. J. (2007). Effect of exposure to arsenic on the development of diabetes mellitus. *Reviews on Environmental Health, 22*(2), 113–130.

Ravenscroft, P., Brammer, H., & Richards, K. (2009). *Arsenic Pollution: A Global Synthesis*. Wiley-Blackwell.

Richardson, S. D., Thruston Jr, A. D., Caughran, T. V., & Chen, P. H. (2007). Colloid and disinfection byproduct controls. *Journal - American Water Works Association, 99*(10), 99–112.

Shirin, S., Jamal, A., Emmanouil, C., Singh, V. P., & Yadav, A. K. (2023). Assessment and characterization of waste material used as backfilling in an abandoned mine. International Journal of Coal Preparation and Utilization, 43(8), 1402–1410. https://doi.org/10.1080/19392699.2022.2118259

Smedley, P. L., & Kinniburgh, D. G. (2002). A review of the source, behaviour and distribution of arsenic in natural waters. *Applied Geochemistry, 17*(5), 517–568.

Smith, A. H., Lingas, E. O., & Rahman, M. (2000). Contamination of drinking-water by arsenic in Bangladesh: a public health emergency. *Bulletin of the World Health Organization, 78*(9), 1093–1103.

Sorg, T. J., Chen, A. S. C., & Westerhoff, P. (2014). Drinking water treatment technology unit cost models. *Journal - American Water Works Association, 106*(2), E91–E101.

Vahter, M. (2008). Health effects of early life exposure to arsenic. *Basic & Clinical Pharmacology & Toxicology, 102*(2), 204–211.

Welch, A. H., Westjohn, D. B., Helsel, D. R., & Wanty, R. B. (2000). Arsenic in ground water of the United States: occurrence and geochemistry. *Ground Water, 38*(4), 589–604.

World Health Organization. (2018). *Arsenic*. Retrieved from https://www.who.int/news-room/fact-sheets/detail/arsenic

Yadav, A. K., Sahoo, S. K., Mahapatra, S., Kumar, A. V., Pandey, G., Lenka, P., & Tripathi, R. M. (2014). Concentrations of uranium in drinking water and cumulative, age-dependent radiation doses in four districts of Uttar Pradesh, India. Toxicological & Environmental Chemistry, 96(2), 192–200. https://doi.org/10.1080/02772248.2014.934247

Chapter 12
Removal of Arsenic from Water Bodies: Current State of Art and Future Perspectives

C. S. Viji and S. Anisha

Abstract Arsenic is one of the most serious pollutants in the environment that mainly pollutes water bodies. It comes from natural sources such as weathering reactions, volcanic eruptions, geothermal reactions, etc., as well as anthropogenic sources such as mining, combustion of fossil fuels, pesticides, etc. The presence of arsenic in drinking water at levels higher than those permitted by the World Health Organization can lead to chronic and fatal diseases like cancer, peripheral neuropathy, renal damage, cardiovascular disorders, etc. Arsenic in any form, organic or inorganic, has a very negative impact on human health though the degree of toxicity varies. Despite the many innovations with regard to the removal of arsenic from water bodies, the exhaustive removal of arsenic remains a challenge. The most commonly used conventional methods of arsenic removal include chemical precipitation, ion exchange, electrocoagulation, nanoparticle-based removal, bioremediation, etc. Although such methods are being implemented in practice, there is always a need to look for cost-effective measures since many of the countries facing a serious threat from arsenic come from the developing or underdeveloped category. In recent times there has been a surge in the number of reports citing cost-effective and environment-friendly methods for treating arsenic-contaminated waters. The methods include the use of Fe-based metal–organic framework, biocomposite filters with nanocrystalline iron oxides in lignocellulose scaffolds, use of microbial consortia, phytoremediation supported by endophytic microbes, use of hydrochar with nanoparticles, etc. Most of these strategies are clever blends of existing procedures that show synergistic effects. However, the variable forms of arsenic and the range of environmental components, especially water bodies, to be ridden off this contaminant leads to challenges in universal implementation of most of the methods. The intelligent application of these technologies can help to better dealing with the problem of arsenic contamination.

C. S. Viji
Department of Fisheries and Aquaculture, St. Albert's College (Autonomous), Ernakulam, India

S. Anisha (✉)
Department of Botany, St. Albert's College (Autonomous), Ernakulam, India
e-mail: sanisha.pillai@gmail.com; anishas@alberts.edu.in

A. K. Yadav et al. (eds.), *Arsenic Removal Technologies*,
https://doi.org/10.1007/978-3-031-91171-2_12

Keywords Organic metal framework · Biocomposite · Nanoparticle · Electrocoagulation · Bioremediation · Toxicity

12.1 Introduction

In today's world, declining environmental quality has exposed human life to numerous threats. Essential resources such as food, water, and air have increasingly become unsafe. The prevalence of incurable diseases underscores the daily dangers facing people. Harmful chemicals are widespread and exposure through ingestion, inhalation, or contact can have severe health impacts. Arsenic, a toxic element, existing as arsenic (III) and arsenic (V) ions, has significantly worsened living conditions in areas with high concentrations in soil, water, and, to a lesser extent, air, the source of which has been attributed to both natural processes such as weathering, volcanic eruptions, geothermal activities, etc., and human-induced activities such as mining, combustion of fossil fuels, etc.

The World Health Organization (WHO 2017) recommends a safe level of 0.01 mg/L of arsenic in drinking water, which, when exceeded for longer periods, is known to be carcinogenic in addition to causing serious diseases such as Blackfoot disease, skin pigmentation, keratosis and nausea (Sarkar and Paul 2016).

To mitigate the health risks related to arsenic in drinking water, it is crucial to develop methods to prevent groundwater contamination (Rahidul Hassan 2023). This article reviews conventional arsenic removal technologies, including chemical precipitation, ion exchange, electrocoagulation, nanoparticle-based methods, and bioremediation. The journal also updates on recent advancements in arsenic removal techniques, such as using iron-based metal–organic frameworks, biocomposite filters with nanocrystalline iron oxides embedded in lignocellulose scaffolds, microbial consortiums, phytoremediation enhanced by endophytic microbes, and hydrochar combined with nanoparticles. Furthermore, the chapter assesses the limitations and potential of various treatment methods and identifies areas for future research and development.

12.2 Complications Associated with Arsenic Removal

The complications associated with conventional arsenic removal are attributed to different factors, including the oxidized state of arsenic, arsenite (As(III)), and arsenate (As(V)), which requires separate procedures for the removal of different forms of arsenic because of the difference in their physicochemical characteristics. Effective removal could be achieved only by the adoption of specific removal technologies for each oxidized state of arsenic. Apart from arsenic, water does contain other anions, such as carbonate, sulfate, phosphate, silicate, and natural organic matter, which could compete with arsenic for available binding sites during the adsorption process

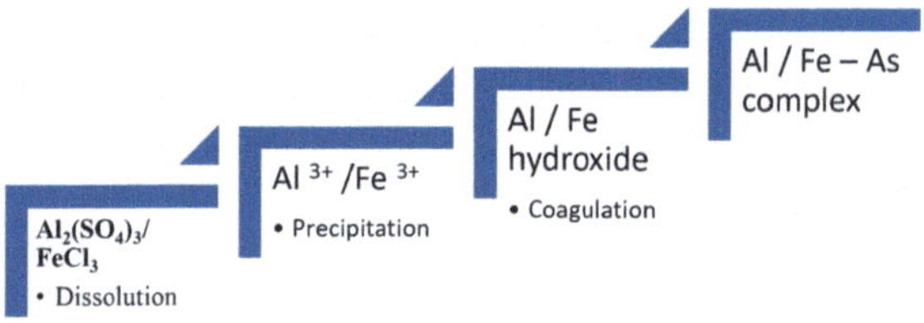

Fig. 12.1 Stages of chemical precipitation

and limit the removal of arsenic (Prateeksha and Swain 2023). pH is yet another factor of concern that is known to affect arsenic removal, and a slightly alkaline pH of 6–8 favors its efficient removal. However, the pH of natural bodies of water usually does not fall within this narrow range, thus limiting the efficiency of arsenic removal (Arslan et al. 2017). In addition to these factors, the reusability of the matrix, scalability, and economic viability also pose complications in the effective removal of arsenic.

12.3 Conventional Methods of Removal of Arsenic

12.3.1 Chemical Precipitation

Chemical precipitation, also known as coagulation, remains the most common method used for the removal of arsenic from water that concentrates on an oxidation pretreatment prior to the process of coagulation and finally combined with filtration and adsorption in (usually) Fe-based media (Fig. 12.1). Studies conducted by Chen et al. (2018) have shown that an oxidation pretreatment using potassium permanganate could improve the removal of arsenic up to 91–92%. Chemical precipitation employs alum for the conversion of toxic components of arsenic into low-solubility compounds, which are later removed by the precipitation process (Alka et al. 2021).

Although the process of chemical precipitation proves to be economically feasible, its effectiveness depends on the pH of the water and the nature of the alum employed for the same. Despite the benefits accompanied with the process, the generation of by-products, flavor, and odour compounds during floc disposal and posttreatment poses serious limitations to the technology (Rahidul Hassan 2023; Shirin and Yadav 2014; Simon et al. 2024).

12.3.2 Ion Exchange

Ion exchange is one of the physicochemical methods used to remove heavy metals like arsenic from polluted water. The process (Fig. 12.2) involves the exchange of chloride or other anions at active sites held electrostatically in the resin with similarly charged ions from the solution (Mondal et al. 2021). The contaminated water, which when flowing through the resins, selectively exchanges ions with those in the resins. Natural and synthetic resins can be used. These resins consist of a three-dimensional hydrocarbon network with ionizable groups that exchange with ions of the same charge in the water. Specifically, strong-base anion exchange resins are effective for removing arsenate because they carry a negative charge, which matches the functional groups that attract and bind to anions. The efficiency of the ion exchange process depends on the presence of the competitive ions present in the contaminated water, which compete with arsenic for the active sites of the resin. This can affect the efficiency of arsenic removal, which in turn requires adjustments in the operation of the ion exchange system.

While ion exchange is effective for the removal of arsenic under suitable conditions, including low total dissolved solids (TDS) and sulfate levels, careful consideration of the water chemistry and resin characteristics is essential to ensure optimal performance of the treatment system. The efficiency of the process is known to improve by incorporating proper adjustments in the operation process based on the nature of the water to be treated (Sarkar and Paul 2016). After use, the exhausted resin, which contains concentrated arsenic, requires further treatment before disposal or potential reuse, ensuring environmental safety and regulatory compliance.

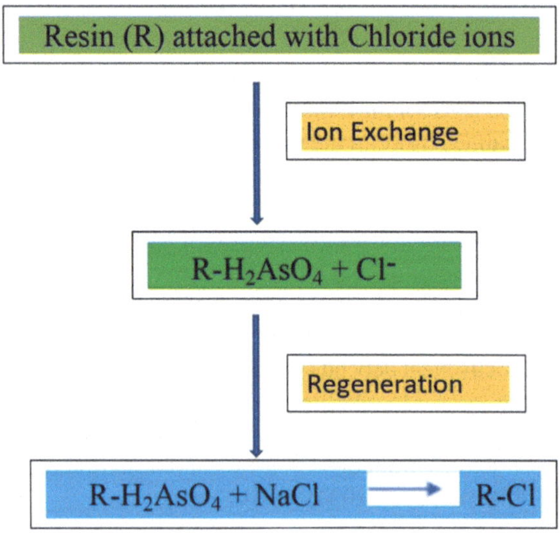

Fig. 12.2 Schematic representation of the ion exchange process

12.3.3 Adsorption

Adsorption emerges as a highly effective technique for the removal of arsenic (As) from water, notable for its efficiency, cost-effectiveness, and ease of implementation, especially in regions with limited resources and infrastructure. Unlike methods requiring chemical additives or a continuous power supply, adsorption is based on the physical attraction between the surface of solid materials (adsorbents) and arsenic ions in water, the intensity of which depends on the contact time, temperature, pH level, competing ions, the amount of adsorbent, etc. (Sarkar and Paul 2016). These variables dictate the capacity to bind arsenic ions effectively, highlighting the need for optimization of specific water chemistry and environmental conditions.

A wide variety of adsorbents are used today, including engineered biochar, iron-based materials, sand coated with metal oxides, activated carbon, activated alumina, etc., for arsenic remediation (Alka et al. 2021). These materials offer various mechanisms for arsenic binding and removal of arsenic, ranging from physical adsorption to chemical reactions with specific surface groups.

The selection of adsorbents depends on their surface area, pore structure, and functional groups, which determine their affinity and capacity for arsenic adsorption. The rate of adsorption is highly dependent on the presence of the competing ions in water apart from arsenic that could compete with arsenic for the binding sites of the adsorbent. Mariko et al. (2022), in their study, investigated the continuous removal of arsenic by adsorption onto a novel modified biosorbent known as iron-coated cork granules (ICG).

Although adsorption remains a pivotal technique for removing arsenic from water because of its efficiency, operational simplicity, economic feasibility, and minimal waste generation, continued advances in adsorbent technology and process optimization are essential for the successful reduction of arsenic levels in contaminated water.

12.3.4 Electrocoagulation

Electrocoagulation (EC) processes are emerging technologies for arsenic from polluted water by employing diverse metal anodes such as aluminum, iron, titanium, etc. During electrocoagulation, the metal anodes generate metal cations, and the cathode liberates hydrogen gas along with the release of hydroxyl radicals in the presence of an electric current. These cations favor coagulation by forming monomeric and polymeric hydroxyl metallic complexes. Comparative studies (Prateeksha and Swain 2023) on chemical coagulation and electrocoagulation in the removal of the different forms of arsenic demonstrated the superior efficiency of electrocoagulation in removing arsenite, while both methods showed similar effectiveness in removing arsenate. Electrocoagulation (EC) has additional drawbacks, including energy consumption and effluent treatment technologies.

12.3.5 Nanoparticle-Based Removal

Nanotechnology, a vital technology for treating contaminated water, employs nanoparticles (iron-based, titanium-based, and carbon-based nanoparticles, etc.), which due to their high adsorption capacity, small size, and large specific surface area, as well as their enhanced adsorption efficiency and potential for functionalization, are ideal for water treatment.

The nanoparticles of iron, nanoscale zero-valent iron (nZVI), were found to be much more effective in converting the pollutants to their less toxic forms through a combined process of adsorption and co-precipitation. Bashir et al. (2022) in their studies have shown that ZrO_2–Fe_3O_4 nanoparticles were capable of converting arsenic into its less toxic form through the process of oxidation in the presence of UV light. At pH 6.5, calcium peroxide (CaO_2) nanoparticles were found to achieve a 91% As (III) removal rate, demonstrating its effectiveness in reducing arsenic concentration to safer levels (Rahidul Hassan 2023).

Hydrous titanium dioxide nanoparticles ($TiO_2 \times H_2O$) are another promising adsorbent that is known to effectively adsorb As(III), eliminating the need for preoxidation and any pH adjustments (Rahidul Hassan 2023). Additionally, Fe-Ti binary oxide nanomaterials, combining the preoxidation and adsorption of both components respectively have shown promise in arsenic removal (Ashraf et al. 2019).

Carbon nanotubes (CNTs), a prominent choice for adsorption, have proven effective in adsorbing various toxic chemicals and pollutants in water, the efficiency of which is based on their large specific surface area, low mass density, high purity, and strong interactions with pollutants (Kindi et al. 2024).

Graphene oxide (GO) nanocomposites are now emerging as an effective component for the removal of arsenic through the process of nanotechnology. Goethite nanoparticles have been cited as effective adsorbents to remove As(V) at an optimal pH of 3.0 (Nicomel et al. 2015). Although nanotechnology proves to be highly efficient in the removal of arsenic because of its enhanced adsorption efficiency, the combustion process of nanoparticles can volatilize arsenic, releasing it into the atmosphere, posing significant environmental and health risks.

12.3.6 Bioremediation

Bioremediation, a new technology that encompasses biological methods, is known to effectively remove arsenic from contaminated water. A notable bioremediation approach is phytoremediation, which utilizes plants to remove pollutants in an environmentally friendly manner. This technique employs plants with extensive root systems and therefore requires minimal nutrient input, which in turn is easily manageable (Rahidul Hassan 2023). Ferns and high-biomass aquatic plants such as *Ceratophyllum*, *Hydrilla*, *Lemna*, *Azolla*, *Pistia*, *Salvinia*, and *Eichhornia* are particularly effective in this regard (Vipin et al. 2024).

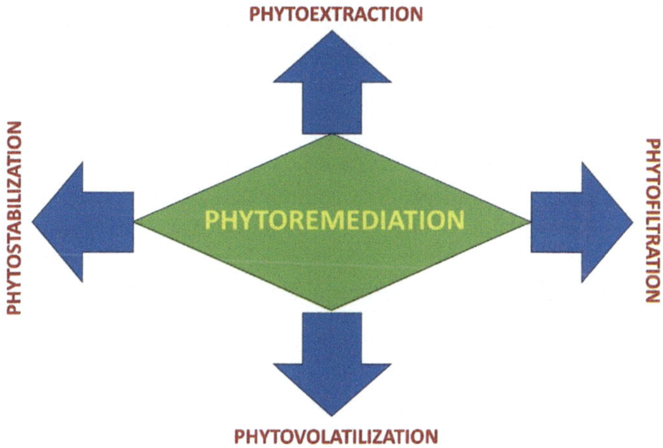

Fig. 12.3 Mechanisms of phytoremediation

Hyperaccumulator plants play a crucial role in the extraction of metals from the soil. Some species transport metals using specific transporter proteins, while others absorb metals directly from aquatic environments through water uptake, leading to adsorption and accumulation in the shoot system of the plant. The process of phytoremediation occurs through various mechanisms (Fig. 12.3), namely phytoextraction (metals removed into plant biomass), phytostabilization (reduces metal mobility through plant activities), phytofiltration (adsorption of metal products by plant parts), and phytovolatilization (toxic metals converted into gaseous forms released into the atmosphere by leaves) (Alka et al. 2021).

In addition to plants, bioremediation also employs sulfate-reducing bacteria and other microorganisms (*Paenibacillus*, *Pseudomonas*, *Haemophilus*, *Micrococcus*, *Bacillus*, etc.) and dissimilatory arsenate-reducing bacteria (ARD) (*Geospirillum arsenophilus*, *Desulfotomaculum auripigmentum* and *Chrysiogenes arsenatis*) for effective removal of arsenic from contaminated water (Rahidul Hassan 2023). However, many arsenic removal techniques do not address neutral As(III) molecules (Mohanty 2017). Consequently, oxidation is often used as a preliminary treatment step, employing different chemical agents. However, this process could produce hazardous by-products.

Nanophytoremediation (NP) is a promising advancement that combines nanomaterials with plants to tackle contaminants after knowing the impacts of the nanomaterials. NP merges phytoremediation with nanotechnology, offering a green and eco-friendly solution for remediation. Nanotechnology introduces novel materials that can improve treatment efficiency, shorten cleanup times, and detoxify contaminated soil and water (Alka et al. 2021). Phytobial remediation, which combines plants and microbes, has been extensively studied for arsenic removal, improving detoxification, stress tolerance, and nutrient availability while promoting beneficial microbial activity in the rhizosphere (Alka et al. 2020).

12.4 Recent Advances in Arsenic Removal Technologies: Efficient, Low-Cost Measures

Arsenic has continued to be a significant cause of concern in the context of the quality of water. The release of arsenic from various natural processes and its multifaceted chemical nature make its removal more complicated. Almost every now and then strategies come up with claims of effective elimination of arsenic. However, whatever the method, few concerns should be kept in mind owing to the complex nature of this contaminant before proceeding (Rahidul Hassan 2023).

(a) In different global locations, there is considerable difference in the amount of As present in water,
(b) Maintaining the pH value of water within a proper range is extremely necessary for the purification process to be effective,
(c) The presence of other ionic species in the water sample significantly affects the efficiency of As removal,
(d) Any method of removal of As should be adopted after considering the social, economic, and technological aspects,
(e) The management of large amounts of hazardous garbage is difficult,
(f) Efficient technological support is mandatory for consistently effective As removal in the long run.

The choice of the arsenic removal method to be used is mainly guided by the predominant state of arsenic oxidation along with the water sample pH. Owing to the serious health concerns associated with the use of arsenic-contaminated water, innovations that lead to efficient removal of arsenic have always been looked forward to.

12.4.1 Biocomposite Filters with Nanocrystalline Iron Oxides in Lignocellulose Scaffolds

Lignocellulose is an abundantly available natural material with great stability which when present in wood, can serve as a filter due to the porous nature of the latter. The porous network of wood can be practically useful as a filter by some appropriate chemical or thermal procedures. Wood gives the advantage of being nature-based and ecofriendly which ensures satisfying the sustainability criterion. Such wood-based filters have been used effectively for the remediation of organic pollutants and crude oil. The use of lignocellulose can help in effective remediation even for nanoscopic particles, including pathogens, at affordable prices.

As a step further towards enhancing the effectiveness of remediation using lignocellulose, nanocrystalline particles have been impregnated in the lignocellulosic pores by precipitation. Balsa wood was subjected to a chemical reaction of iron salts and potassium hydroxide to produce the precipitation of magnetite and ferrihydrite in a nanocrystalline form inside the lignocellulosic network. The matrix or filter so obtained could be standardized for its specifications like grain size, nanostructure, etc., depending upon the different types of contaminants. Since the impregnated nanocrystalline compounds were rather minimally soluble, there was very little leaching out of the same and it was available for long periods for the remediation procedures. Arsenic in both oxidation states, As(III) and As(V), was effectively taken up by both forms of iron, magnetite, and ferrihydrite, at neutral pH (Soini et al. 2024).

Because wood having varied properties is widely available, it can be suitably used properly for the effective removal of arsenic. This can be a very meaningful utilization of the enormous biomass available as wood for the purification of water, especially for drinking purposes. The minimal interventions needed from the technological side and the low resource input requirements make it a very attractive and efficient option for arsenic removal mainly for nations that have limitations in the above aspects. The lignocellulose thus can be effectively used with less financial trouble and almost no environmental problems for the remediation of arsenic-contaminated waters. There will be a need to customize the form of lignocellulose and wood to be used in different settings by different nations for arsenic removal.

12.4.2 Metal–Organic Frameworks

Metal–organic frameworks (MOFs) or MOCs are versatile and efficient molecular configuration where metal ions serve as centers of coordination and facilitate the linking of various ligands. The inherent structure of MOFs is such that they tend to have a large surface area. Moreover, their porous nature imparts considerable adsorptive potential to MOFs. MOFs and MOF-based materials (mMOFs) have been efficiently used for removing arsenic from water and have evolved considerably (Nguyen et al. 2023). Their efficiency further increases by their self-assembly into high-grade permanent structures with enhanced surface area and porosity. Although with some negatives such as toxicity and costly production procedures, MOFs and MOF-based materials (mMOFs) are highly efficient in remediation and show great versatility of usage. They can be fabricated with other materials or integrated with membranes rendering them more efficient, stable, and reusable. Future approaches to improving will involve reducing the production costs and environmentally friendly disposal (Nguyen et al. 2023).

12.4.3 Use of Microbial Consortia

Microorganisms have shown varying degrees of tolerance to metalloids, and this potential is more common in the species present in regions exposed to accumulation of such substances. Tolerance and the ability to take up such hazardous substances make them an ideal choice for bioremediation.

Bioremediation is a preferred approach for cleaning up the environment since it is an environmentally friendly method that is not only not very expensive but also permits the safe removal of toxicity and the recovery of certain metalloids for industrial purposes. Though bioremediation primarily involves bacteria, it may include fungi, algae or plants, etc., under the names mycoremediation, phytoremediation, etc., respectively. Bioremediation aims to reduce the presence of contaminant substances such as arsenic in the environment and also to reduce their toxicity and adverse effects (Lashani et al. 2023). Microorganisms have various strategies for the removal of arsenic from contaminated sites and also for the reduction of the toxicity or hazard associated with the metalloid in the environment (Table 12.1). Bacteria collected from arsenic-contaminated sites and having the potential to oxidize arsenic include *Bacillus arsenooxydans*, *Alcaligenes faecalis*, *Pseudomonas arsenitoxidans*, *Microbacterium lacticum*, *Agrobacterium tumefaciens*, *M. oxydans*, *P. stutzeri*, *Aeromonas* sp., *Agrobacterium* sp., *Comamonas* sp., *Enterobacter* sp., *Brevundimonas*, *Acidovorax*, *Acinetobacter*, *Undibacterium*, *Herbaspirillum*, *Rhodococcus*, *Staphylococcus*, *Bosea*, *Ralstonia*, *Caulobacter*, *Pantoea* sp., *Pseudomonas* sp. and *P. lubricans* (Sher and Rehman 2019; Lashani et al. 2023). Not only bacteria but certain fungi are also resistant to arsenic and can reduce the same. These include members from *Aspergillus*, *Trichoderma*, *Sordaria*, *Neocosmospora*, etc.

A wide range of genes responsible for arsenic-related metabolic mechanisms, mainly from bacteria, have been characterized. These include genes that facilitate metalloid uptake (GlpF), its ionic conversions by microbes (both within and outside the cell), impart resistance or tolerance to the microbes towards arsenic (arsABC operon), and elimination of arsenic from the cell (Lashani et al. 2023).

Although the literature gives a very promising picture of the bioremediation approach to solving the issue of arsenic pollution, in many cases, the microbial systems may not be able to reach their maximum remedial efficiencies. Various factors play a significant role in deciding the degree of removal of arsenic from microbe-contaminated sites (Fig. 12.4).

Table 12.1 Microbial mechanisms of removing arsenic from the environment and reducing its toxicity

Sl. No.	Mechanisms of removal of arsenic	Mechanisms of decreasing toxicity caused by arsenic
1	Bioaccumulation	Change in chemical properties of arsenic
2	Biosorption	Change in oxidation state
3	Biotransformation	Enhancing solubility of arsenic
4	Bioleaching	Immobilization by precipitation
5	Biomineralization	
6	Microbially enhanced chemisorption	

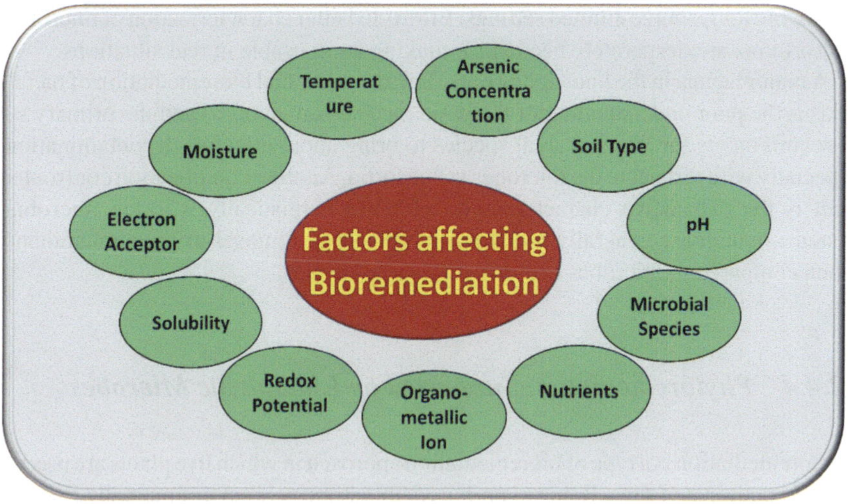

Fig. 12.4 Factors affecting the degree of bioremediation brought about by microbes

Many studies on the bioremediation of metalloids, including arsenic, have been conducted, but most of them involve a single microorganism. However, in real situations, the contaminated sites host a range of microbes, either as a consortium or as a biofilm. It is the microbial ecosystem of the consortia or the biofilm that actually is able to bring about the removal of arsenic. Biofilms have been established to be better at removing environmental contaminants than single species (Lashani et al. 2023). Microbial consortiums are much more efficient in bioremediation due to the synchronous application of the potentials of the various microbial species present in the consortia. Synergistic interactions between the members of the consortiums are very crucial in scaling up the quantum of bioremediation. Putting such consortiums to practical applications calls for a proper and exhaustive characterization. However, characterizing such a multitude of microbes as a system is a huge challenge since knowledge of aspects like the component members, structure of the microbial ecosystem, their interactions, metabolic needs, and the mechanism for all these. The interactions among the members in a consortium can be positive, negative, or neutral. The most effective and productive interactions involve cooperation in which the action of one species of microbes is beneficial to the other by helping in its growth or by removal or breakdown of antagonistic substances or by supply of growth factors, etc. (Lashani et al. 2023). Effective consortia have been shown to exist between bacteria and fungal members. This consortium of *Bacillus*, *Sphingomonas* (bacteria), and *Fusarium* (Fungus) was characterized for its simultaneous removal of arsenic and pyrene (Lashani et al. 2023). Microbial consortiums are much more efficient in reducing the problem of arsenic contamination and can be applied in practical application to remediate contaminated sites. They are both sustainable and affordable for resource-limited settings. Efforts to better characterization of microbial consortiums are desperately needed for making them usable in real situations.

A major lacuna in the knowledge regarding the microbial bioremediation of metalloids is the poor understanding of the basic mechanism involved and the primary set of requirements for the microbial species to bring about effective decontamination, especially with respect to the microbes in consortia. Another complication confronted with is the exhaustive characterization of the synergistically working microbial systems, which are generally much more efficient in removal of the contaminants when compared to microbes occurring as single species.

12.4.4 Phytoremediation Supported by Endophytic Microbes

Phytoremediation is a type of bioremediation approach in which live plants are used to clean contaminated sites. Being plant-based, this strategy is environmentally friendly and also cost-effective but is largely compromised with regard to the pace of environmental cleanup. There can also be the problem of phytotoxicity. Among plants, three key characteristics are mainly considered desirable for arsenic phytoremediation: (i) Fast growth; (ii) high accumulation of arsenic and (iii) appreciable biomass. However, all three features are rarely available in one plant (Srivastava et al. 2021). *Pteris vittata* has been studied extensively due to its remarkable hyperaccumulation potential for arsenic, and it has also been used in the field for remediation purposes. Although the maximum arsenic uptake by the plant is appreciable and also comparable to other conventional methods for arsenic removal, it suffers from a lack of consistency that leads to variable performances (Yang et al. 2020).

A new strategy is to use the microorganisms associated with plants to enhance the arsenic remediation efficiency by improving the three required features listed above (Srivastava et al. 2021). Not only rhizosphere microorganisms, but also endophytes are able to improve phytoremediation efficiency by their direct interactions within plants.

The endophytes maintain a symbiosis within plants in which they receive nutrition and space to live, while in return they provide enhanced resistance to plants against biotic and abiotic stress. Arsenic imposes abiotic stress on plants which is relieved by endophytes that are resistant to these substances and in many cases they breakdown these contaminants, thus effecting phytoremediation (Nguyen and Phan 2022).

Arsenic induces oxidative stress in plants and plants respond by producing defense molecules such as phytochelatins (PCs), non-protein thiols (NPTs), cysteine, and glutathione (GSH) for detoxification. The role of microbes, both internal and external, in deciding the kind of plant response in arsenic-contaminated sites is extremely significant. Plant–microbe interactions decide the fate of arsenic. Some mechanisms of microbes that influence phytoremediation include microbial production of plant growth factors such as indole-3-aceticacid (IAA), production of organic acids and soil acidification, production of siderophores causing metal chelation, methylation, dissolution of metal phosphates, etc. The bacteria also synthesise 1-amino-cyclopropane-1-carboxylic acid deaminase (ACCD) which soothes the arsenic induced stress in plants. In this regard, the bacteria in the rhizosphere help with the mobility and uptake of the metalloid while the endophytes help improve plant growth by the release of growth hormones and important vitamins. Studies have been carried out to characterize the plant microbial systems involved in phytoremediation and their efficiency of arsenic removal. The plant *Betula celtiberica*, which grows in arsenic-contaminated soils, was investigated for its associated microorganisms and the system was analyzed for its phytoremediation potential based on the microbial profile. The study clearly indicated a strong dependence of the efficiency of phytoremediation on the nature of plant–microbe interaction (Mesa et al. 2017).

12.4.5 Use of Hydrochar and Nanoparticles

Hydrochar is a carbon-rich material generally created by hydrothermal carbonization (HTC), which is a technology involving thermochemical conversion that makes use of moderate temperature and autogenic pressure for synthesis in a liquid medium (Li et al. 2022). Hydrochars are considered better than biochars because of the presence of a large number of functional groups compared to the latter, which supports better metal adsorption. However, they suffer from a poor porous structure and limited surface area that have a significant bearing on their efficiency. Iron modification is an approach usually followed in biochar to improve texture, and the same has been tested to enhance the porosity of hydrochar. The use of iron compounds during hydrochar synthesis has yielded positive results in the form of hydrochars with better adsorption of heavy metals. In addition, it was also revealed that the base material for the hydrochar and the specific HTC conditions had a significant impact on the final product formed. Iron modification was a promising method to enhance the porosity and adsorption sites. In one of the most impactful attempts, the iron retention rate of hydrochar was significantly improved by the use of $Fe(NO_3)_3{\cdot}9H_2O$. Not only this, the arsenic uptake efficiency of the resultant product was also found to be much better than before (Zhang et al. 2024).

It is a well-known fact that nanoparticles are characterized by a high surface-to-volume ratio which imparts considerable adsorptive capacity to these entities. Hydrochars have also been activated using iron-based nanoparticles, which led to the formation of a hydrochar with exceptional adsorption capacity for arsenic. The hydrochar used livestock waste as the starting material and gave a fantastic end product (Chen et al. 2021). Similar approaches can be followed by making use of similar materials such as farm waste and other organic substances that do not have any apparent use for making adsorbent material like hydrochar. This can be a very cost-effective approach for poor countries. Although nanoparticles have been used for remediation by adsorption, these systems face the challenge of maintaining nanoparticles as separate units without aggregation, which is extremely necessary for getting advantage of the huge surface area. Additionally, the aggregation of nanoparticles leads to a reduction in the stability of the adsorbent and also in its function. The required energy inputs also increase to reduce the size of the particles (Nguyen et al. 2023). Addressing the issue of aggregation can prove to be phenomenal in enhancing the efficiency of nanoparticle-based arsenic removal methods based on nanoparticles.

12.5 Conclusion

Arsenic is a serious contaminant in the environment and technologies are desperately needed to decontaminate affected sites are desperately needed. Although many strategies have been tested and standardized for the same, the efficiency of the procedures, availability of the required technological advancement and necessary resources, and many other factors bring about huge variations in the results. In recent times a number of simple and cost-effective measures have been tried for arsenic removal. Significant strategies include the ones based on natural products such as bioremediation, biochar, hydrochar, etc. The common remarkable feature of these products is their sustainable and ecofriendly nature along with minimal financial inputs required. Strategies may be refined with novel innovative interventions to provide cost-effective methods for arsenic removal.

Acknowledgements The authors would like to express their sincere thanks to St. Albert's College (Autonomous), Ernakulam for all the support rendered.

References

Alka, S., S. Shahir, N. Ibrahim, T. Chai, Z. M. Bahari and F. A. Manan, 2020. The role of plant growth promoting bacteria on arsenic removal: A review of existing perspectives. *Environmental Technology and Innovation*. 17: 100602. https://doi.org/10.1016/j.eti.2020.100602

Alka, S., S. Shahir, N. Ibrahim, M. J. Ndejiko, D. N. Vo and F. A. Manan, 2021. Arsenic removal technologies and future trends: A mini review. *Journal of Cleaner Production* 278: 123805. https://doi.org/10.1016/j.jclepro.2020.123805

Arslan, A., L. A. Richards and, P. Bhattacharya, 2017. Arsenic Remediation of Drinking Water: An Overview. In *Best Practice Guide on the Control of Arsenic in Drinking Water*, edited by P. Bhattacharya, D. A. Polya, and Dragana Jovanovic. IWA Publishing 79–98. https://doi.org/10.2166/9781780404929

Ashraf, S., A. Siddiqa, S. Shahida, and S. Qaisar, 2019. Titanium-based nanocomposite materials for arsenic removal from water: A review. *Heliyon* 5(5). https://doi.org/10.1016/j.heliyon.2019.e01577

Bashir, H., I. Bibi, M. M. Hussain, N. K. Niazi, and J. Iqbal, 2022. Nanoparticulate iron oxide minerals for arsenic removal from contaminated water. In *Global Arsenic Hazard: Ecotoxicology and Remediation* (pp. 459–479). Cham: Springer International Publishing. https://doi.org/10.1007/978-3-031-16360-9_21

Chen, H., J. Xu, H. Lin, X. Zhao, J. Shang, and Z. Liu, 2021. Arsenic removal via a novel hydrochar from livestock waste co-activated with thiourea and γ-Fe_2O_3 nanoparticles. *Journal of Hazardous Materials, 419*, 126457. https://doi.org/10.1016/j.jhazmat.2021.126457

Chen, S. C., L. Wang, D. A. Lytle, and T. J. Sorg, 2018. Arsenic/Iron Removal From Groundwater With Elevated Ammonia and Natural Organic Matter. *American Water Works Association, 110*(3), E2–E17. https://doi.org/10.5942/jawwa.2018.110.0020

Kindi, A. L., U. M. Kakar, Uzair Mohammad, Z. Tauanov, S. Azat, N. A. Ahmadi, M. H. & Hassand, A. Sarwari, A. W. Monib, P. Niazi, 2024. Carbon Nanotubes as Adsorbents for Heavy Metals: Focus on Arsenic and Hydrargyrum Removal from Water. *Journal of Chemistry Studies* 3, 07–20. https://doi.org/10.32996/jcs.2024.3.1.2

Lashani, E., M. A. Amoozegar, R. J. Turner, and H. Moghimi, 2023. Use of microbial consortia in bioremediation of metalloid polluted environments. *Microorganisms, 11*(4), 891. https://doi.org/10.3390/microorganisms11040891

Li, J., L. Li, M. Suvarna, L. Pan, M. Tabatabaei, Y. S. Ok, and X. Wang, 2022. Wet wastes to bioenergy and biochar: A critical review with future perspectives. *Sci. Total Environ.* 817, 152921. https://doi.org/10.1016/j.scitotenv.2022.152921

Mariko, A. C., A. M. A. Pintor, R. A.R. Boaventura and C.M.S. Botelho, 2022. Efficient removal of arsenic from aqueous solution by continuous adsorption onto iron-coated cork granulates. *Journal of Hazardous Materials* 432: 128657, https://doi.org/10.1016/j.jhazmat.2022.128657

Mesa V., R. Navazas AGonzález-Gil, A. González, N. Weyens, B. Lauga, J. L. R. Gallego, J. Sánchez, and A. I. Peláez, 2017. Use of Endophytic and Rhizosphere Bacteria To Improve Phytoremediation of Arsenic-Contaminated Industrial Soils by Autochthonous Betula celtiberica. Appl Environ Microbio l83: e03411-16. https://doi.org/10.1128/AEM.03411-16

Mohanty, D., 2017. Conventional as well as emerging arsenic removal technologies-a critical review. *Water Air and Soil Pollution* 228 (10):1–21. https://doi.org/10.1007/s11270-017-3549-4

Mondal, P., P. K. Roy, N. S. Mondal, S. K. Basak, A. Majumder, 2021. Development of Low-Cost Arsenic Removal Process by Using Ion-Exchange Resins. In *Advances in Water Resources Management for Sustainable Use*, edited by P. K. Roy, M. B. Roy, M.B., Pal, S. Lecture Notes in Civil Engineering (131). Springer 139–147. https://doi.org/10.1007/978-981-33-6412-7_12

Nguyen, D. A., D. V. Nguyen, G. Jeong, N. Asghar, and A. Jang, 2023. Critical evaluation of hybrid metal–organic framework composites for efficient treatment of arsenic-contaminated solutions by adsorption and membrane-separation process. *Chemical Engineering Journal, 461*, 141789. https://doi.org/10.1016/j.cej.2023.141789

Nguyen, K. B. T., and T. H. T. Phan, 2022. October. Application of plant endophytic microorganisms (Endophytes) in the treatment of heavy metal pollution in Soils. In *International Conference on Geo-Spatial Technologies and Earth Resources* (pp. 199–221). Cham: Springer International Publishing. https://doi.org/10.1007/978-3-031-17808-5_13

Nicomel, N. R., K. Leus, K. Folens, V. D. P. Voort and D. G. Laing, 2015. Technologies for Arsenic Removal from Water: Current Status and Future Perspectives. *International Journal of Environmental Research* and *Public Health* 22;13(1):ijerph13010062. https://doi.org/10.3390/ijerph13010062

Prateeksha, M. and R. Swain, 2023. A mini-review on arsenic remediation techniques from water and future trends. *Water Science and Technology* 87 (12): 3108–3123. https://doi.org/10.2166/wst.2023.190

Rahidul Hassan, H, 2023. A review on different arsenic removal techniques used for decontamination of drinking water. *Environmental Pollutants and Bioavailability, 35* (1), 2165964. https://doi.org/10.1080/26395940.2023.2165964

Sarkar, A. and B. Paul, 2016. The global menace of arsenic and its conventional remediation—A critical review. *Chemosphere* 158:37–49. https://doi.org/10.1016/j.chemosphere.2016.05.043

Shei, S., and Rehman, A. 2019. Use of heavy metals resistant bacteria-a strategy for arsenic bioremediation. *Applied microbiology and biotechnology, 103* (15), 6007–6021. https://doi.org/10.1007/s00253-019-09933-6

Shirin, S., & Yadav, A. K. (2014). Physico-chemical analysis of municipal wastewater discharge in Ganga river, Haridwar district of Uttarakhand, India. Current World Environment, 9(2), 536–543. https://doi.org/10.12944/CWE.9.2.39

Simon, M., Joshi, H., Yadav, A. K., & Hopke, P. K. (2024). Simultaneous nitrogen and phosphorus removal from polluted water using a novel bacterial consortium (BACON-38): Promising application for wastewater remediation. Journal of Water Process Engineering, 66, 106044. https://doi.org/10.1016/j.jwpe.2024.106044

Soini, S. A, S. M. Feliciano, B. G. Duersch, and V. M. Merk, 2024. Nanocrystalline iron hydroxide lignocellulose filters for arsenate remediation. *RSC sustainability*, *2*(3), 626–634. https://doi.org/10.1039/D3SU00326D

Srivastava, S., A. Shukla, V. D. Rajput, K. Kumar, T. Minkina, S. Mandzhieva, A. Shmaraeva, and P. Suprasanna, 2021. Arsenic Remediation through Sustainable Phytoremediation Approaches. Minerals, 11, 936. https://doi.org/10.3390/min11090936

Vipin, C. J., S. Sharma and A. Bhattacharya, 2024. Intertwining of materials into de-arsenification of water. *Bulletin of Materials Science* 47:83. https://doi.org/10.1007/s12034-024-03178-4

WHO, 2017. Guidelines for drinking-water quality, 4th edition incorporating the first addendum. Geneva, *World Health Organization* 315–318. https://apps.who.int/iris/bitstream/handle/10665/254637/9789241549950-eng.pdf

Yang, C., Y. N. Ho, C. Inoue, and M. F. Chien, 2020. Long-term effectiveness of microbe-assisted arsenic phytoremediation by Pteris vittata in field trials. *Science of the Total Environment*, *740*, 140137. https://doi.org/10.1016/j.scitotenv.2020.140137

Zhang, Y., D. Chen, Y. Xing, B. Liu, Y. Zhou, and P. Lu, 2024. Role of nitrate in the production of iron-modified hydrochar for arsenic removal. *Water, Air, & Soil Pollution*, *235*(4), 233. https://doi.org/10.1007/s11270-024-07023-6

Chapter 13
Arsenic Pollution Mitigation Through Technological Developments in Developing Countries

Divya Patel, Ananthalakshmi Ranganathan, Anoop Kumar Kanaujia, and Seema Paroha

Abstract Two areas severely affected by arsenic pollution are Bangladesh and West Bengal of India, while some studies indicated that approximately 108 countries are affected by groundwater contamination with arsenic. Bangladesh is reportedly the world's worst arsenic disaster, with over half of its population at risk of arsenic toxicity. It is critical to mitigate arsenic pollution through technologies developed in arsenic-affected regions. Financial and social limits pose significant obstacles to the use of technology to mitigate arsenic pollution mitigation. Many arsenic removal techniques, such as membrane filtration, coagulation, and adsorption, are technically needed for complex infrastructure and maintenance, which may not be possible in environments with limited resources. The financial burden of implementing and maintaining these technologies can be unaffordable, especially in developing nations such as Bangladesh. Adopting new technology is sometimes hampered by local ideas and customs, which can also be important but frequently overlooked factors. These include lack of awareness, sociocultural factors, community acceptability, and involvement. However, there are enough chances for innovation and advancement. Technological developments that are affordable and easy to utilize present attractive options for reducing arsenic exposure. Novel adsorbents, solar oxidation, biosand filters, etc. are some examples of emerging techniques that have the potential to remove arsenic in a way that is more accessible and sustainable. To improve acceptability and long-term success, there is also a growing understanding of the importance of including community education and participation in technology adoption techniques. This chapter emphasizes the necessity for a multidisciplinary strategy that integrates technological innovation, economic viability, and sociocultural sensitivity while focusing on the possibilities and limitations of implementing arsenic technology.

D. Patel · A. Ranganathan (✉) · A. K. Kanaujia · S. Paroha
National Sugar Institute, Kanpur, India
e-mail: ananthuknocks@gmail.com

D. Patel
Department of Environmental Studies, MSU, Baroda, Vadodara, India

A. K. Yadav et al. (eds.), *Arsenic Removal Technologies*,
https://doi.org/10.1007/978-3-031-91171-2_13

Keywords Arsenic contamination · Drinking water · Remediation · Innovative technologies · Challenges and opportunities

13.1 Introduction

The long history of groundwater and drinking water contamination by arsenic. The rationale for controlling the amount of arsenic in drinking water comes from the well-established effects of long-term exposure to the element (Council and Do 1999; Levine 1987). Millions of people worldwide are affected by drinking water contaminated with arsenic, especially in South and Southeast Asia such as Bangladesh, India, Vietnam, Nepal, and China (Shaji et al. 2021). Although arsenic comes from geogenic sources, human activities may accelerate its mobility and concentration in groundwater. Examples of these activities include arsenic-treated lumber, fracking, coal-fired power plants, mining, and arsenic-containing pesticides (McCarty et al. 2011; Shirin and Yadav 2021). Acute exposure can cause symptoms including diarrhoea, nausea, and vomiting. On the contrary, chronic exposure has been associated with major health problems such as skin lesions, heart problems, and several types of cancer (Hughes et al. 2011). Both inorganic and organic forms of arsenic exist, and each has particular impacts on human health. Because of its toxicity, inorganic arsenic is especially hazardous. The most frequent way that humans are exposed to arsenic is through contaminated drinking water. Waters that come into contact with specific rocks and soils may naturally contain arsenic due to its occurrence. Products made of food and drink might expose people. Natural sources of drinking water often contain arsenic levels as low as 1–2 μg/L, except for areas afflicted by arsenic (Bhat et al. 2024). Authorities have implemented numerous recommendations to resolve the issues caused by arsenic contamination; however, some requirements must be met in order to proceed: The suggested limit for this type of water is 10 mg/L (Ahmad and Bhattacharya 2019). Effective treatment frameworks must be developed, suitable supply sources must be identified, and clear considerations must be selected. Understanding the sources and distribution channels of arsenic pollution is necessary to develop effective mitigation strategies. The main ways that people are exposed to inorganic arsenic are by drinking water and other food items.

This chapter aims to evaluate the various arsenic mitigation strategies and interventions implemented globally from the perspective of sustainability, emphasizing the variables that affect their viability and, ultimately, suggesting an approach to strategy. The primary focus is on arsenic mitigation strategies and the key obstacles facing arsenic mitigation technologies.

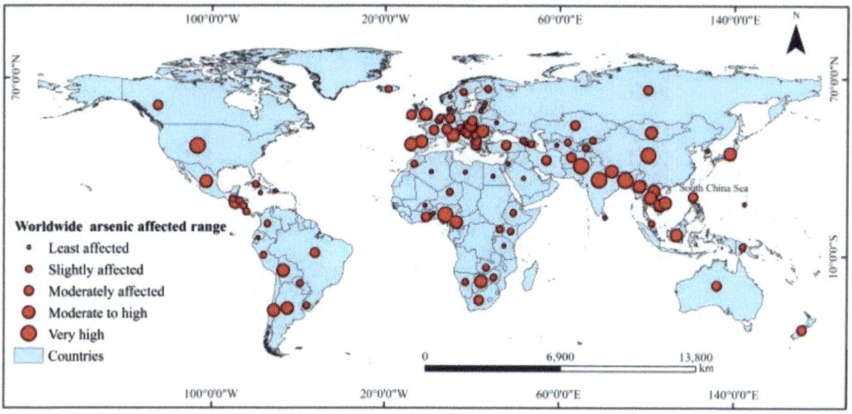

Fig. 13.1 Regions impacted by arsenic worldwide. Reprinted from Shaji et al. (2021) with permission from Geoscience Frontiers. Copyright 2021, Elsevier Publishing

13.2 Global Prevalence of Arsenic Contamination

The natural arsenic pollution of groundwater has been documented globally, as shown in Fig. 13.1, with most cases occurring in South American and South Asian locations (Shaji et al. 2021). However, the regions of Bangladesh, India, Nepal, Vietnam, and China that together make up the South and Southeast Asian Belt are considered to be the most arsenic-polluted (McCarty et al. 2011). In Asian countries, groundwater serves as the primary drinking water supply. However, the previous study revealed that Southeast Asia's groundwater contains significant levels of naturally occurring and man-made arsenic. Consequently, >80 million people are susceptible to arsenic exposure (Rahman et al. 2023). About 110 million people in Asian nations drink water that is rich in arsenic (Table 13.1). In Bangladesh, almost 20 million people are exposed to water with arsenic concentrations higher than 50 μg/L, and as of right now, more than 3 million of the 11 million wells in the nation have arsenic levels higher than 10 μg/L (Raessler 2018). Up to 4730 μg/L of arsenic has been detected in groundwater in India (Chakraborti et al. 2018). To protect future generations, extra care must be taken because, in many places, the amount of arsenic in groundwater remains unknown despite extensive studies.

13.3 Health Impact of Arsenic

Arsenic is the most toxic metal that may be found in nature. Not from mines, smelting or agricultural sources, but rather from natural geological pollution of drinking water, is the main cause of arsenic poisoning in people. Arsenic exposure can occur by inhalation, skin absorption, and consumption of contaminated drinking water (Rajiv

Table 13.1 Population at risk due to arsenic contamination

Countries	Population at risk (lakhs)	Refs.
Bangladesh	750	Natasha et al. (2020)
Pakistan	500–600	
India	500	
China	196	
Vietnam	50	
Myanmar	25	
Cambodia	24	
Nepal	22.9	
Taiwan	0.1	
Argentina	20	Shaji et al. (2021)
Mexico	20	
Chile	5	
Spain	5	
France	0.17	

et al. 2023). According to studies, arsenic can have a variety of acute or chronic adverse effects on human health, as illustrated in Fig. 13.2. Both carcinogenic and non-carcinogenic effects are probable with arsenic. According to the WHO, arsenic is a Class 1 carcinogen (Demissie et al. 2024).

According to the International Diabetes Federation, arsenic consumption is also associated with Pakistan's high incidence of diabetes. However, no research was conducted in Pakistan to assess the potential relationship between arsenic in the country's groundwater and the high incidence of diabetes (Shahid et al. 2018).

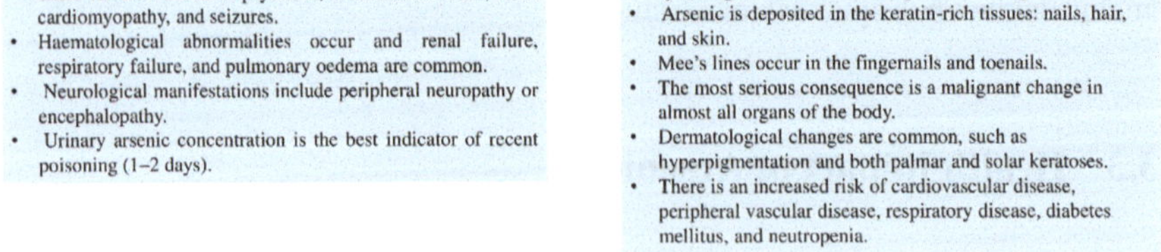

Fig. 13.2 Effect of arsenic toxicity

13.4 Arsenic Remediation Technologies

In recent decades, many studies have been conducted on the removal of arsenic from the environment to enhance arsenic uptake. The researchers are investigating these mechanisms in both field and laboratory environments. Adsorption, chemical precipitation, bioremediation, ion exchange, phytoremediation, phytobial remediation, electrokinetic techniques, and electrocoagulation are the techniques currently being used (Fig. 13.3).

13.4.1 Coagulation/Filtration

The coagulation procedure involves mixing arsenic-contaminated water with metal-based coagulants, such as ferric chloride ($FeCl_3$). In water, $FeCl_3$ hydrolyzes to

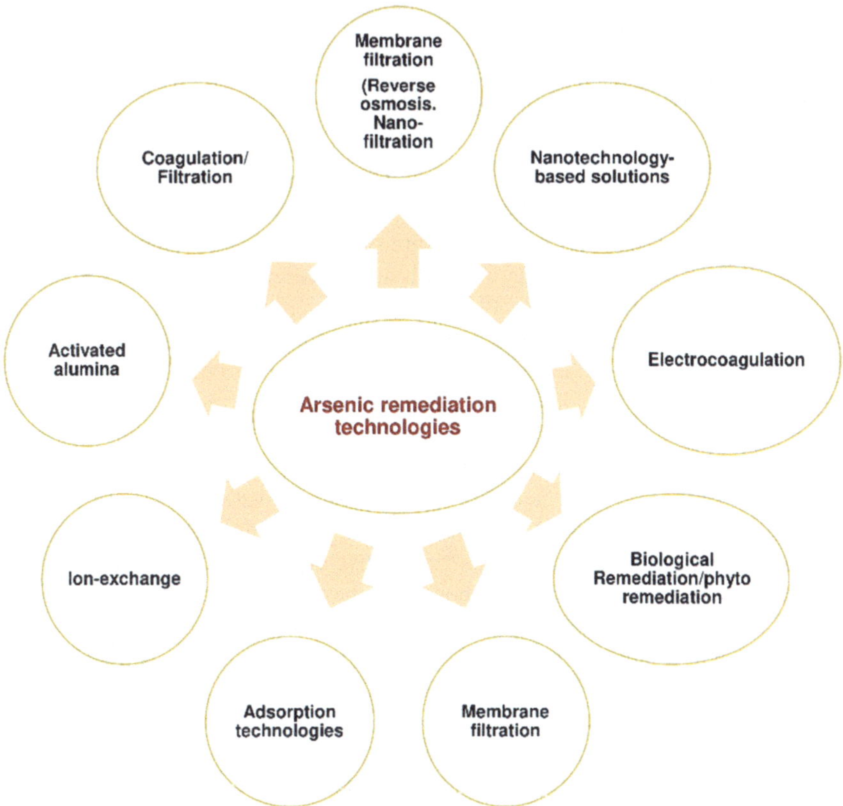

Fig. 13.3 Overview of arsenic remediation technologies

produce ferric hydroxide, which is positively charged ($Fe(OH)_3$). Effective elimination of arsenic requires oxidation [As(V)]. During coagulation, the negative charge of colloids is reduced by ferric chloride ($FeCl_3$) and aluminum sulfate ($Al_2(SO_4)_3$), which causes the particles to collide and expand. However, by adding an anionic flocculant, flocculation occurs by charge neutralization or bridging between the larger particles that form, leading to the formation of flocs. The dissolved arsenic is transformed into an insoluble solid by the chemicals involved in these reactions, and then the solid precipitates.

In 2004, Wickramasinghe and colleagues investigated the feasibility of using coagulation with ferric sulfate and ferric chloride as coagulants, followed by filtration, to remove arsenic from groundwater in Sonargaon, Bangladesh, and a city in southern Colorado, USA. They found that using any of the coagulants would produce a practical solution (Wickramasinghe et al. 2004). The primary drawback of flocculation-coagulation is the large-scale production of concentrated arsenic sludge. This sludge needs to be controlled in order to prevent secondary pollution of the environment. Additionally, sludge remediation is costly. These limitations make this approach less feasible, especially in field circumstances (Nicomel et al. 2015).

13.4.2 Adsorption Technology

Adsorption technology is an effective method for removing a wide range of solutes from water. Adsorption is a viable method for removing arsenic that is both efficient and economically viable. It does not require the addition of chemicals, and it can be easily applied in developing nations with limited access to electricity and skilled labor. It has been shown to have >95% quantitative efficiency for the remediation of arsenite and arsenate (Kumar et al. 2019). This depends on the van der Waals separation process and the electrostatic forces of attraction between the adsorbed molecules. However, it should be mentioned that temperature, initial arsenic concentration, adsorbent dose, pH, duration of exposure, and the presence of other chemical species determine how effective this approach is (Samimi et al. 2022). Arsenic can be effectively removed from aqueous solutions using a variety of adsorbents that have been found so far; few are tabulated in Table 13.2. Although a lot of emphasis has recently been placed on the use of innovative engineered materials to remove As. Graphene oxide, metal oxides, carbon nanotubes, and metal–organic frameworks (MOFs), biochars, have been researched for this reason (Dehghani et al. 2023).

The primary variables that affect arsenic adsorption in continuous mode are the amount of adsorbent (bed height), adsorbate initial concentration of the adsorbate, pH, temperature, flow rate, hydraulic retention duration, and the presence of other chemical components. According to the findings of previous research, iron inclusion is the most widely used technique for boosting the arsenic adsorption capacity of adsorbent materials, followed by the addition of other metals such as aluminium and zirconium (Yadav et al. 2022). To achieve greater adsorption, several kinds of materials have been investigated and improved. It is discovered that several adsorbents

Table 13.2 List of various adsorbents that have been published along with a comparison analysis

Adsorbents	Optimum pH	Sorption capacity (mg/g)	Refs.
Coconut-shell carbon	5.0	2.40	Nicomel et al. (2015)
Coconut-shell carbon pretreated with Fe(III)	5.0	4.53	
Coal-based carbon	5.0	4.09	
Rice polish	7.0	0.15	
Fly ash	4.0	30	
Eggshell membrane	7.0	24.2	
Synthetic zeolite H-MFI-24	6.5	35.8	
Granular titanium dioxide	7.0	41.4	
Granular ferric hydroxide (GFH)	6.5	1.1	
Iron-oxide-coated manganese sand (IOCMS)	7.0	5.452	
Iron-modified activated carbon	7.6–8.0	51.3	
Amorphous iron hydroxide	6.0–8.0	7.0	
Zero-valent iron	10.0	1.92	
Goethite	6.0–8.0	4.0	
$Fe_x(OH)_y$-montmorillonite	6.0–8.0	4.0	
Ti_xH_y-montmorillonite	6.0–8.0	3.0	
Natural siderite	7.31	0.52	
Kaolinite	5.0	0.86	
Modified calcined bauxite	7.0	1.566	
Activated red mud	7.25/3.50	0.941	
Cerium-loaded cation exchange resin	5.0–6.0	1.03	
Surface-modified diatomite	7.0	8.0	
Fungal strains	6.0	90.0	Jaiswal et al. (2018)
Magnetic biochar (MBC) from corn straw-derived biochar	4.0	22.94	Khan et al. (2020)
Iron-coated cork granulates	3	4.2	Carneiro et al. (2022)

have a rather good adsorption capacity. However, the majority of the research was preliminary and there is still a gap between the development of new adsorbents and feasibility assessments for real-world use.

13.4.3 Membrane Filtration

Membrane filtration is a potential technique for eliminating all heavy metals and TDS from water, in addition to eliminating arsenic from it. In this regard, membrane technology is considered one of the most promising. Membranes are man-made materials with billions of pores or microscopic holes that act as barriers, allowing some substances to pass through, but preventing others. The pressure differential between the two sides of the membrane drives the molecules through the membrane during the membrane filtration process. There are different types of membrane processes, shown in Fig. 13.4. Since the eighteenth century, membranes have been established. Since then, numerous advancements have been made to improve the suitability of membranes for a wide range of applications (Yin et al. 2024). The oldest and finest technique for helping modest water treatment systems remove arsenic from water is reverse osmosis. According to Dutta et al., RO membranes have incredibly tiny pores (<0.001 μm), which allow for a very high (sometimes close to 100%) rejection of ions and molecules (Dutta et al. 2012). Ultrafiltration (UF) is essentially a pressure-driven membrane separation technology based on size exclusion. UF membranes may hold species with molecular weights between 300 and 500,000 Da and typically have pore diameters between 10 and 10,000 Å. Polymers, colloidal particles, carbohydrates, and biomolecules are examples of species that are typically rejected. UF membrane technology is a unique physicochemical approach developed by TRL to successfully remove arsenic from water sources. The effectiveness of this technology is bringing the exposure to a safe level. This technique offers a comprehensive solution for water purification by efficiently eliminating excessive amounts of iron, manganese, bacteria, and other substances in addition to arsenic (TRL 2024). The larger pore size of current pristine ultrafiltration (UF) membranes, which generally range between 0.02 and 0.1 μm, makes it difficult for them to remove arsenic from water. However, in some studies, UF membranes are designed to separate arsenic ions from the water matrix in a variety of ways. Arsenic can be eliminated by combining polymers with various metal- and carbon-based nanoparticles (Sánchez and Rivas 2010). Micellar enhanced ultrafiltration (MEUF) for an arsenic-contaminated aqueous solution was optimized by Yaqub et al. using cetylpyridinium chloride (CPC) as a surfactant. Experiments and modeling with artificial neural networks (ANNs) were used to study this process. As a result of its great computational speed, accuracy, and economy in MEUF process optimization—all without the need for laborious experimental work—they concluded that the ANN model can be considered a competitive, potent, and efficient replacement (Yaqub and Lee 2021).

Microfiltration (MF) is a low-pressure membrane technique that is used to separate suspended and colloidal particles of sizes between 0.1 and 5 μm. It is very similar to traditional coarse filtration or sieving methods. True solution molecules can pass through this kind of membrane due to its high porosity. However, arsenic-contaminated water cannot have its dissolved arsenate and arsenite species removed by microfiltration alone. However, increasing the size of the arsenic-containing particle is the only way to eliminate arsenic using an MF membrane. Before MF,

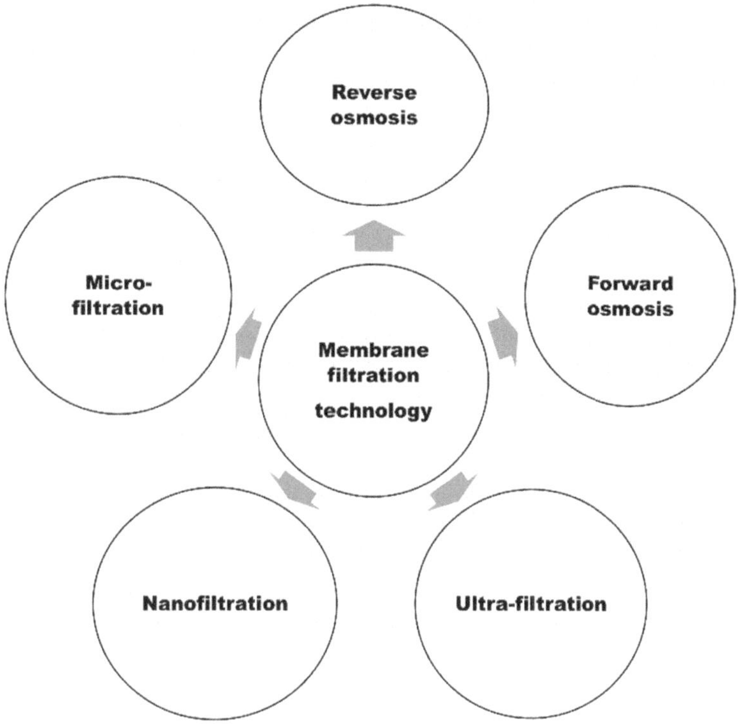

Fig. 13.4 Membrane filtration and its type

it was demonstrated that coagulation and flocculation procedures could effectively increase the size of arsenic-bearing species' particles and remove arsenic species from water contaminated with arsenic.

Ghurye et al. (2004). Nanofiltration, with its pores 0.001–0.01 μm-sized pores, is one of the most modern and efficient pressure-based membrane techniques to separate multivalent ions from monovalent ones (Birniwa et al. 2024). Most commercial NF membranes have a stable surface charge because of surface groups (such as sulfonated or carboxyl acids) dissociating from one another. This enables ion separation by combining ion interaction processes, pore size, and electrical effect (Chabi Noel and Chen 2021). In order to extract water from dissolved solutes, forward osmosis (FO), another method of purifying water, uses a semipermeable membrane and the osmotic pressure's natural energy. Numerous investigations have been conducted to enhance the possibility of eliminating arsenic. To treat arsenic-containing water, Yang and his colleagues, for instance, developed and synthesized several novel imidazolium-based ionic liquids using one-step quaternization processes. They further grafted these novel compounds onto conventional thin-film composite FO membranes (Yang et al. 2019).

Bioremediation/phytoremediation Bioremediation, which employs microorganisms as sorbents (microbial remediation) or plants as green factories (phytoremediation) as green factories, is one of the methods for eliminating arsenic from water that is considered to be both cost-effective and ecologically safe. Microbes can stabilize and detoxify heavy metals, reducing their impact on the environment, through enzymatic reduction reactions to oxidoreductases. Acting mainly on As(III) and As(V) through redox reactions, as-resistant bacteria isolated from a variety of environmental samples convert inorganic to organic forms. To detoxify or accumulate As, a variety of metabolic processes are involved, such as oxidation, reduction, biosorption, methylation, and volatilization (Irshad et al. 2021). Dey et al. used 16S rRNA sequencing and biochemical analysis to identify *Bacillus* sp. and *Aneurinibacillus aneurinilyticus*. They found that 51.45, 51.99, and 53.29% of arsenite could be converted to less harmful arsenate by isolated SW2 and SW4. Two isolates (SW2 and SW4) showed growth that grew with increasing arsenate concentration to 3000 ppm, then gradually reduced and finally stopped growing at 4500 ppm of arsenate, according to the optical density (Dey et al. 2016). In addition to bacteria, other microorganisms such as fungi, algae, etc. (Table 13.3). have also been employed in arsenic bioremediation from soil and water (Irshad et al. 2021). Another practical and attractive approach to decontaminating arsenic-contaminated soil and water is phytoremediation (Table 13.3), while the proper methods and possibly beneficial plants are used. *Lemna valdiviana* Phil., an aquatic plant species, was used by Souza et al. to extract arsenic from an aqueous solution. They discovered that Lemna valdiviana Phil absorbed more arsenic when cultured at pH values between 6.3 and 7.0, with a readily available phosphorus ($P–PO_4$) concentration of 0.0488 mmol L^{-1} and nitrogen in the form of 7.9 mmol L^{-1} nitrate. Under these conditions, the plants managed to absorb 1190 mg kg^{-1} As (in dry weight) from the aqueous medium and reduce its original concentration by 82% (Souza et al. 2019).

13.5 Challenges in Implementing Arsenic Technologies

Several challenges to overcome when implementing arsenic removal methods, especially in drinking water treatment, such as technical challenges, economic challenges, institutional or policy challenges, environmental challenges, health challenges, etc. The activated alumina-assisted method had a cost per m^3 of water ranging from $2.39 to US $3.20, but despite its exceptional efficacy in removing arsenic, it was not a viable economic means of treating arsenic-contaminated water, particularly in developing and underdeveloped nations (Kabir and Chowdhury 2017). Arsenic contamination with arsenic is a serious problem, particularly in low and lower-middle-income nations where the average daily income is $2.01–10 and US $10.01–20, respectively (Kabir and Chowdhury 2017). Considering that there are so many impoverished people living in these countries' rural areas, it is unlikely that they will have the money to use advanced arsenic removal treatment methods (Kabir and Chowdhury

Table 13.3 List of few microorganisms and plants as remediation of arsenic

Sl. No.	Microorganisms/plants/ferns	Accumulation/tolerance	Refs.
1	*Pityrogramma calomelanos* (leaf-bearing spore)	8000 μg/g (accumulation)	Visoottiviseth et al. (2002)
2	*Reynoutria sachalinensis*	1900 μg/g (accumulation)	
3	*Wolffia globosa*	>1000 mg/kg (accumulation)	Zhang et al. (2009)
4	*Aspergillus* sp.	10,000 mg/L (tolerance)	Srivastava et al. (2011)
5	*Ceratophyllum demersum* L.	963 μg/g (accumulation)	Xue et al. (2012)
6	Strain of *Aeromonas*	100 mM arsenate (tolerance)	Abbas et al. (2014)
7	*Bacillus* sp.	4500 mg/L (tolerance)	Dey et al. (2016)
8	*Aneurinibacillus aneurinilyticus*		
9	*Azolla caroliniana*	386.1 μg/g (accumulation)	Leão et al. (2017)
10	*Hydrilla verticillata*	197.2 ± 17.4 μg/g (accumulation)	Zhao et al. (2020)
11	*Lantana camara*	–	Preetha et al. (2023)
12	*Betula celtiberica*	–	
13	*Eichhornia crassipes*	498.4 mg/kg (accumulation)	
14	*Landolita punctata*	>1000 g/kg (accumulation)	

"–" Data not available

2017). The main challenges to the techniques developed to treat arsenic are listed in Table 13.4.

The majority of adsorbents in Arsenic adsorption technology are pH-dependent, so maintaining the solution's pH before treatment poses another challenge. The implementation of arsenic removal technologies faces numerous social and cultural challenges. Some of the key challenges include lack of awareness, misinformation, cultural perception, and trust in technology (Singha 2021). The inconsistent use of technology for removing arsenic can result from weak regulatory frameworks. Attempts to remedy arsenic contamination can be hampered in some places by a lack of enforcement of water safety regulations (Bhat et al. 2024).

Table 13.4 Major challenges associated with the methods developed for arsenic treatment

Arsenic treatment methods	Primary challenges
Ion exchange	Expensive medium. It requires high-tech maintenance and operation. An issue with sludge disposal results from regeneration. As(III) is challenging to eliminate. Resin regeneration is worn out. Useful for low-TDS. Resin's lifespan
Chemical precipitation	Generates poisonous sludge, it might be necessary to pre-oxidize, which can lead to dangerous disinfection byproducts. principally eliminates As(V) and low removal of As(III), oxidation, sedimentation, and filtering of chemicals are required
Membrane technique	Maintenance and operation costs are high. Hazardous wastewater may be produced. Pre-treatment is required
Adsorption	The absorption bed becomes worn out. Adsorbent material replenishment regularly. It creates solid hazardous waste
Electrocoagulation	A strong foundation is required. Effective design and operating criteria must be the focus
Phytoremediation	A strong foundation is required. Not readily available in stores. Very slow and seasonally effective treatment method
Bioremediation	Slow as compared to chemical treatment. Incomplete removal of particularly in environments with high concentrations of arsenic, limited by environmental conditions such as pH, temperature, oxygen, etc.

13.5.1 *Opportunities in the Implementation of Arsenic Technologies*

The use of arsenic remediation technologies offers considerable opportunities, particularly in places with acute contamination problems. Arsenic in drinking water and soil poses health risks that can be reduced by research and innovative technological developments in adsorption, membrane filtration, and bioremediation, which are scalable and affordable processes. These technologies can be customized to meet different environmental conditions and resource availability, enabling their implementation in both urban and rural contexts. Additionally, the development of locally produced, low-cost materials for arsenic removal, such as agricultural by-products or modified natural minerals, might increase accessibility and sustainability. By generating jobs and supporting sectors of the economy devoted to public health and environmental protection, the incorporation of these technologies into water management systems can also promote economic growth. Furthermore, the application of arsenic technologies promotes global efforts to address one of the most urgent environmental health concerns by providing opportunities for international collaboration, knowledge exchange and innovation. Arsenic-based projects can be customized to match the unique requirements and cultural contexts of affected populations by incorporating local groups in the development and execution of these technologies. This ensures greater acceptability and long-term use. Encouraging local populations to participate in educational programs on the risks posed by arsenic and the advantages of remediation methods can encourage people to actively participate in protecting their water sources and promoting clean water. Moreover, utilizing community resources and expertise can result in creative, locally tailored, economical, and environmentally friendly solutions.

13.6 Conclusion and Recommendations

In this chapter, we cover a variety of promising study innovations and future research options about the identification and mitigation of arsenic exposures and their health effects are covered in this chapter. The pros and cons of arsenic treatment systems are also discussed in this chapter. Implementing arsenic remediation technologies presents numerous challenges but also offers significant opportunities to improve public health and environmental sustainability. By addressing technical, economic, social, and policy-related challenges, and leveraging opportunities for innovation and community engagement, it is possible to make significant progress in providing safe drinking water to affected populations. Although a great deal of highly advanced research and development has been done in the field of treating arsenic, many countries are making extensive use of these technologies. Despite these developments in arsenic technologies, many people, particularly in developing and underdeveloped nations, cannot use them due to their high cost and limited availability. Therefore, it is recommended that robust, low maintenance systems be designed in order to treat arsenic-contaminated water while keeping these flaws in mind. Create basic monitoring tools and integrate training programs for local operators to guarantee the technology's continued functionality.

Acknowledgements The authors thank our prestigious institute and other people who are involved in this study for their direct and indirect contributions.

References

Abbas SZ, Riaz M, Ramzan N, Zahid MT, Shakoori FR, Rafatullah M. Isolation and characterization of arsenic resistant bacteria from wastewater. Braz J Microbiol 2014; 45: 1309–15.

Ahmad A, Bhattacharya P. Arsenic in Drinking Water: Is 10 μg/L a Safe Limit? Current Pollution Reports 2019; 5.

Bhat A, Ravi K, Tian F, Singh B. Arsenic Contamination Needs Serious Attention: An Opinion and Global Scenario. Pollutants 2024; 4: 196-211.

Birniwa AH, Habibu S, Abdullahi SSa, Mohammad REA, Hussaini A, Magaji H, et al. Membrane technologies for heavy metals removal from water and wastewater: A mini review. Case Studies in Chemical and Environmental Engineering 2024; 9: 100538.

Carneiro MA, Pintor AMA, Boaventura RAR, Botelho CMS. Efficient removal of arsenic from aqueous solution by continuous adsorption onto iron-coated cork granulates. Journal of Hazardous Materials 2022; 432: 128657.

Chabi Noel W, Chen Z-L, Bacharou T. Arsenic removal from water by nanofiltration membrane: Potentials and limitations. Water Practice and Technology 2021; 16.

Chakraborti D, Singh S, Rahman MM, Dutta R, Mukherjee S, Pati S, et al. Groundwater Arsenic Contamination in the Ganga River Basin: A Future Health Danger. International Journal of Environmental Research and Public Health 2018; 15: 180.

Council NR, Earth Do, Studies L, Sciences CoL, Water SoAiD. Arsenic in drinking water. 1999.

Dehghani MH, Ahmadi S, Ghosh S, Othmani A, Osagie C, Meskini M, et al. Recent advances on sustainable adsorbents for the remediation of noxious pollutants from water and wastewater: A critical review. Arabian Journal of Chemistry 2023; 16: 105303.

Demissie S, Mekonen S, Awoke T, Teshome B, Mengistie B. Examining carcinogenic and noncarcinogenic health risks related to arsenic exposure in Ethiopia: A longitudinal study. Toxicol Rep 2024; 12: 100–110.

Dey U, Chatterjee S, Mondal NK. Isolation and characterization of arsenic-resistant bacteria and possible application in bioremediation. Biotechnology Reports 2016; 10: 1–7.

Dutta T, Bhattacherjee C, Bhattacherjee S. Removal of arsenic using membrane technology—a review. Int. J. Eng. Res. Technol 2012; 1: 1–23.

Ghurye G, Clifford D, Tripp A. Iron coagulation and direct microfiltration to remove arsenic from groundwater. Journal American Water Works Association - J Amer Water Work Assoc 2004; 96: 143–152.

Hughes MF, Beck BD, Chen Y, Lewis AS, Thomas DJ. Arsenic exposure and toxicology: a historical perspective. Toxicol Sci 2011; 123: 305–32.

Irshad S, Xie Z, Mehmood S, Nawaz A, Ditta A, Mahmood Q. Insights into conventional and recent technologies for arsenic bioremediation: A systematic review. Environmental Science and Pollution Research 2021; 28: 18870–18892.

Jaiswal V, Saxena S, Kaur I, Dubey P, Nand S, Naseem M, et al. Application of four novel fungal strains to remove arsenic from contaminated water in batch and column modes. Journal of Hazardous Materials 2018; 356: 98–107.

Kabir F, Chowdhury S. Arsenic removal methods for drinking water in the developing countries: technological developments and research needs. Environmental Science and Pollution Research 2017; 24: 24102–24120.

Khan ZH, Gao M, Qiu W, Qaswar M, Islam MS, Song Z. The sorbed mechanisms of engineering magnetic biochar composites on arsenic in aqueous solution. Environmental Science and Pollution Research 2020; 27: 41361–41371.

Kumar R, Patel M, Singh P, Bundschuh J, Pittman CU, Trakal L, et al. Emerging technologies for arsenic removal from drinking water in rural and peri-urban areas: Methods, experience from, and options for Latin America. Science of The Total Environment 2019; 694: 133427.

Leão G, Oliveira J, Felipe R, Farnese F. Phytoremediation of arsenic-contaminated water: the role of antioxidant metabolism of Azolla caroliniana Willd. (Salviniales). Acta Botanica Brasilica 2017; 31.

Levine T. Special report on ingested inorganic arsenic: skin cancer; nutritional essentiality: US Environmental Protection Agency, 1987.

McCarty KM, Hanh HT, Kim KW. Arsenic geochemistry and human health in South East Asia. Rev Environ Health 2011; 26: 71–8.

Natasha N, Shahid M, Imran M, Khalid S, Murtaza B, Niazi N, et al. Arsenic Environmental Contamination Status in South Asia, 2020, pp. 13–39.

Nicomel NR, Leus K, Folens K, Van Der Voort P, Du Laing G. Technologies for Arsenic Removal from Water: Current Status and Future Perspectives. Int J Environ Res Public Health 2015; 13: ijerph13010062.

Preetha JSY, Arun M, Vidya N, Kowsalya K, Halka J, Ondrasek G. Biotechnology Advances in Bioremediation of Arsenic: A Review. Molecules 2023; 28: 1474.

Raessler M. The Arsenic Contamination of Drinking and Groundwaters in Bangladesh: Featuring Biogeochemical Aspects and Implications on Public Health. Archives of Environmental Contamination and Toxicology 2018; 75.

Rahman MS, Reza AHMS, Ahsan MA, Siddique MAB. Arsenic in groundwater from Southwest Bangladesh: Sources, water quality, and potential health concern. HydroResearch 2023; 6: 1–15.

Rajiv SV, George M, Nandakumar G. Dermatological manifestations of arsenic exposure. Journal of Skin and Sexually Transmitted Diseases 2023; 5: 14–21.

Samimi M, Zakeri M, Alobaid F, Aghel B. A Brief Review of Recent Results in Arsenic Adsorption Process from Aquatic Environments by Metal-Organic Frameworks: Classification Based on Kinetics, Isotherms and Thermodynamics Behaviors. Nanomaterials (Basel) 2022; 13.

Sánchez J, Rivas BL. Arsenic extraction from aqueous solution: electrochemical oxidation combined with ultrafiltration membranes and water-soluble polymers. Chemical Engineering Journal 2010; 165: 625–632.

Shahid M, Niazi NK, Dumat C, Naidu R, Khalid S, Rahman MM, et al. A meta-analysis of the distribution, sources and health risks of arsenic-contaminated groundwater in Pakistan. Environ Pollut 2018; 242: 307–319.

Shaji E, Santosh M, Sarath KV, Prakash P, Deepchand V, Divya BV. Arsenic contamination of groundwater: A global synopsis with focus on the Indian Peninsula. Geoscience Frontiers 2021; 12: 101079.

Shirin, S., & Yadav, A. K. (2021). Assessment and suitability for the dumping of flyash in an abandoned mine. Environmental Challenges, 4, 100111. https://doi.org/10.1016/j.envc.2021.100111

Singha S. Role of Public Awareness Ensuring the Sustainability of Arsenic Mitigation. Journal of Pharmaceutical Research International 2021; 33: 391–404.

Souza TDd, Borges AC, Braga AF, Veloso RW, Teixeira de Matos A. Phytoremediation of arsenic-contaminated water by Lemna Valdiviana: An optimization study. Chemosphere 2019; 234: 402–408.

Srivastava PK, Vaish A, Dwivedi S, Chakrabarty D, Singh N, Tripathi RD. Biological removal of arsenic pollution by soil fungi. Science of The Total Environment 2011; 409: 2430–2442.

TRL. Arsenic Removal Filter. In: Ltd. TR, editor, 2024.

Visoottiviseth P, Francesconi K, Sridokchan W. The potential of Thai indigenous plant species for the phytoremediation of arsenic contaminated land. Environmental Pollution 2002; 118: 453–461.

Wickramasinghe SR, Han B, Zimbron J, Shen Z, Karim MN. Arsenic removal by coagulation and filtration: comparison of groundwaters from the United States and Bangladesh. Desalination 2004; 169: 231–244.

Xue P, Yan C, Sun G, Luo Z. Arsenic accumulation and speciation in the submerged macrophyte Ceratophyllum demersum L. Environmental Science and Pollution Research 2012; 19: 3969–3976.

Yadav MK, Saidulu D, Ghosal PS, Mukherjee A, Gupta AK. A review on the management of arsenic-laden spent adsorbent: Insights of global practices, process criticality, and sustainable solutions. Environmental Technology & Innovation 2022; 27: 102500.

Yang Q, Lau CH, Ge Q. Novel Ionic Grafts That Enhance Arsenic Removal via Forward Osmosis. ACS Applied Materials & Interfaces 2019; 11: 17828–17835.

Yaqub M, Lee SH. Experimental and neural network modeling of micellar enhanced ultrafiltration for arsenic removal from aqueous solution. Environmental Engineering Research 2021; 26.

Yin H, Xu M, Luo Z, Bi X, Li J, Zhang S, et al. Machine learning for membrane design and discovery. Green Energy & Environment 2024; 9: 54–70.

Zhang X, Zhao F-J, Huang Q, Williams PN, Sun G-X, Zhu Y-G. Arsenic uptake and speciation in the rootless duckweed Wolffia globosa. New Phytologist 2009; 182: 421–428.

Zhao Y, Zhen Z, Wang Z, Zeng L, Yan C. Influence of environmental factors on arsenic accumulation and biotransformation using the aquatic plant species Hydrilla verticillata. Journal of Environmental Sciences 2020; 90: 244–252.

Chapter 14
Building Capacity and Knowledge Sharing for Sustainable Arsenic Management

Ambrina Sardar Khan, Prateek Srivastava, and Anand Mishra

Abstract The presence of arsenic in groundwater poses a significant threat to public health and the environment, particularly in areas that depend on aquifers for drinking water and agriculture. To address this issue, a multidisciplinary approach is necessary that includes capacity building and knowledge sharing for sustainable arsenic management. Capacity building involves developing technical skills and competencies among local stakeholders, such as government officials, water managers, and community leaders, through training programs and workshops focused on arsenic detection, risk assessment, and mitigation strategies. The use of advanced technological tools, such as portable field kits and remote sensing technologies, improves the accuracy and efficiency of arsenic assessment in water sources, enabling real-time monitoring and data-driven decision-making. It is crucial to engage communities, raise awareness of the risks of arsenic exposure, and promote safe water practices. Educational campaigns and participatory approaches can promote community ownership and improve the sustainability of the implemented solutions. Effective policy frameworks and institutional support, including strict guidelines for arsenic levels in drinking water, compliance enforcement, and financial and technical support for mitigation initiatives, are critical to sustainable arsenic management. Knowledge exchange between researchers, policymakers, and practitioners can be facilitated through international conferences, workshops, and online repositories, which can help advance management strategies. The development and implementation of sustainable mitigation technologies, such as adsorption filters, phytoremediation, and microbial bioremediation, are integral to reducing arsenic levels in affected areas. Through multidisciplinary collaboration, technological innovation, community engagement, robust policy frameworks, and continuous knowledge exchange, sustainable solutions can be devised and implemented to mitigate the adverse consequences of arsenic exposure and protect public health.

A. S. Khan (✉) · A. Mishra
Department of Environmental Sciences, Integral University, Lucknow, India
e-mail: ambrina@iul.ac.in; ambiikhan@gmail.com

P. Srivastava
Department of Botany, University of Allahabad, Prayagraj, India

A. K. Yadav et al. (eds.), *Arsenic Removal Technologies*,
https://doi.org/10.1007/978-3-031-91171-2_14

Keywords Arsenic · Groundwater · Capacity building · Community engagement · Policy frameworks · Knowledge sharing · Mitigation strategies

14.1 Introduction

Arsenic, a naturally occurring element, is ubiquitous in the Earth's crust and can be released into groundwater through natural processes and anthropogenic activities. Its presence in drinking water poses a severe threat to public health, leading to various diseases, such as skin lesions, cardiovascular diseases, and cancers. Arsenic groundwater contamination is a global concern, particularly in regions with a high geological arsenic content, such as South Asia. Addressing this challenge requires a multifaceted approach that includes capacity building, technological innovation, community involvement, policy support, and knowledge exchange. Strong leadership and collaboration across sectors are crucial to effectively address this challenge and create sustainable solutions (Sobandi et al. 2021).

The issue of arsenic management in sustainable development has attracted attention in recent years due to its significant potential impact on humans and the planet. Arsenic groundwater contamination is a pressing global issue, particularly in regions such as South Asia. Approximately millions of inhabitants of these regions are exposed in some or other way to arsenic concentrations that exceed the World Health Organization (WHO) guideline value of 10 μg/L. Figure 14.1 shows sustainable arsenic management using various means and technologies.

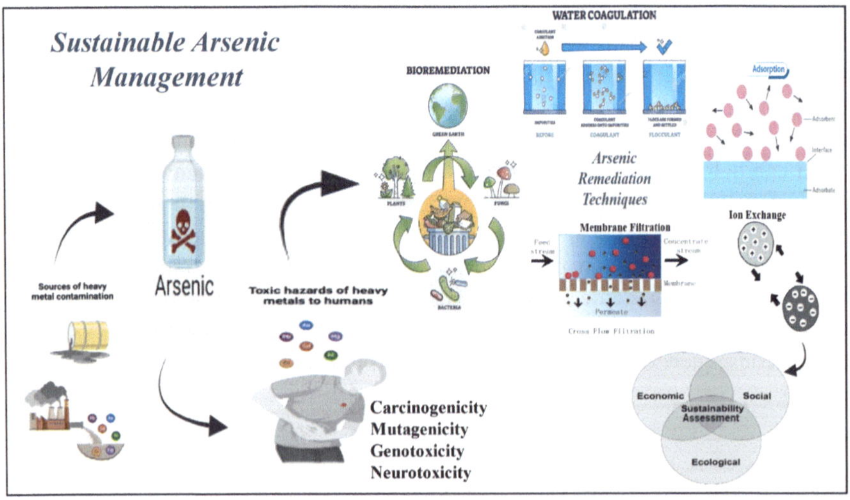

Fig. 14.1 Sustainable arsenic management using various means and technologies

14.2 Background

Arsenic contamination of groundwater is a widespread global issue that affects 21 countries and poses significant health risks, particularly in developing nations, where groundwater is often the preferred drinking source. This problem is especially severe in the Bengal Delta Plain region, including Bangladesh and West Bengal, India, where millions of people are at risk of arsenic poisoning. The biogeochemical cycle of As involves various physicochemical processes and biological mechanisms, with microorganisms playing a key role. The inorganic compounds arsenite [As(III)] and arsenate [As(V)] are the most toxic and abundant species in water, are readily transported through aqueous routes and accumulate in food crops and aquatic plants. This contamination has far-reaching consequences on society and the economy, threatening sustainability in the affected areas, as shown in Fig. 14.2.

Arsenic contamination of groundwater has severe global environmental, social, and economic impacts. The presence of As in water poses a significant threat to human health, causing various cancers, skin lesions, neurological damage, heart disease, and hypertension. Chronic exposure to arsenic also negatively affects early childhood development, impacting physical, socioemotional, and learning skills. Environmentally, arsenic contamination not only affects human health but also endangers the survival and long-term sustenance of nonhuman species. It causes toxicity through oxidative stress, DNA damage, behavioral dysfunction, endocrine disruption, and epigenetic alterations in wildlife. Additionally, the use of arsenic-contaminated water for irrigation degrades soil quality, leading to poor agricultural production and economic losses. Arsenic contamination creates widespread problems for affected individuals and their families. Arsenicosis patients face discrimination and ostracism

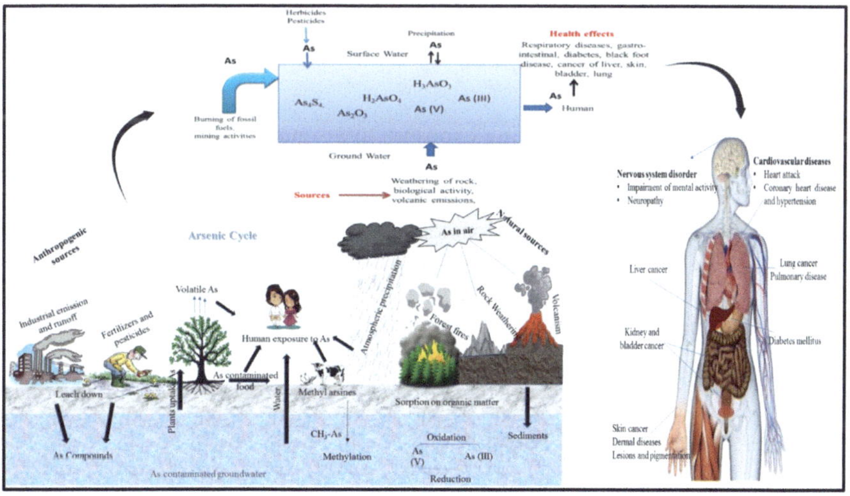

Fig. 14.2 Arsenic contamination and associated problems

within their communities, and many people mistakenly believing it is a contagious disease. Women with visible symptoms struggle to marry, and some affected housewives face divorce. Children with symptoms are often excluded from school to hide the problem. The social stigma associated with arsenic poisoning significantly affects the quality of life of affected individuals. Economically, arsenic contamination places a heavy burden on affected communities, particularly in marginalized sections. A significant portion of earnings (up to 60% in some cases) is spent on hospitalization and treatment. Contamination also affects agricultural productivity, leading to economic losses for farmers. Furthermore, the need for arsenic mitigation technologies and safe drinking water provision limits local and national resources. The social and economic impacts of As contamination were significant. In the Shahpur block of Bihar, India, studies have shown that low educational levels hinder sustainable solutions, and people suffering from arsenic-related diseases face discrimination. Contamination has also impacted food security, drinking water availability, and social cohesion in affected regions.

Arsenic contamination has far-reaching consequences for the environment, society, and economy. It presents a complex challenge with interconnected biological, social, and economic consequences. It not only poses immediate health risks, but also creates long-term challenges for sustainable development in affected areas. Addressing this issue requires a multifaceted approach, including awareness campaigns, sustainable mitigation measures, and policies that consider the complex interplay between environmental, social, and economic factors.

14.3 Overview of the Impact of Arsenic on Human and Environment

14.3.1 Human Health Impact

As shown in Fig. 14.3, long-lasting exposure to arsenic-contaminated water leads to various adverse health effects, including skin warts and blisters, cardiovascular disease, and various forms of cancer, notably skin, bladder, and lung cancer. Arsenic exposure is associated with developmental effects, neurotoxicity, and diabetes (Naujokas et al. 2013). The latency period for some of these diseases can be long, making it difficult to establish direct causal links and necessitating long-term studies to understand the full impact of exposure.

14.3.2 Environmental Impact

Arsenic contamination can disrupt ecosystems. It can accumulate in soils and sediments, thereby affecting agricultural productivity and food safety (Rahman et al.

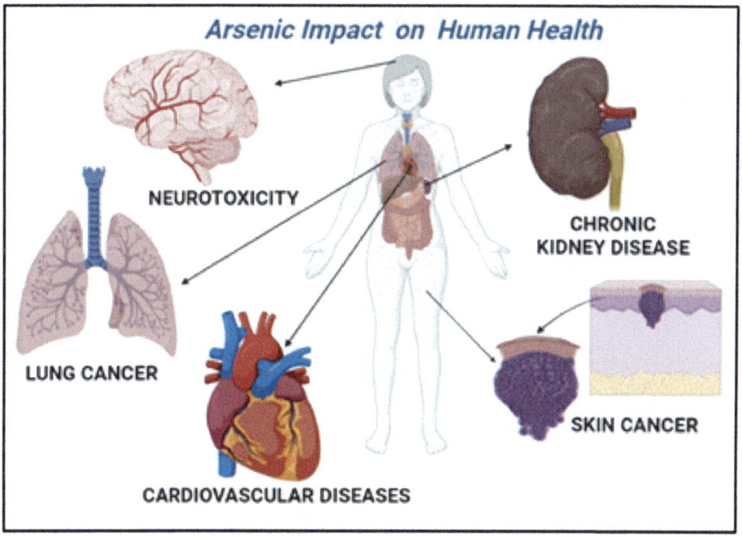

Fig. 14.3 Impact of arsenic on human health

2019). Consumption of crops, such as rice, which absorbs arsenic and serves as a staple food in many affected areas, presents considerable health hazards to individuals throughout the food chain. Furthermore, when surface water becomes contaminated, it can negatively impact aquatic ecosystems, leading to the accumulation of As in fish and other organisms (Sarkar et al. 2004).

14.3.3 Socioeconomic Impact

Arsenic contamination has a substantial socioeconomic impact. Communities impacted by this issue often struggle financially due to decreased productivity caused by health problems and higher healthcare costs (Dasgupta et al. 2015). The availability of clean drinking water is crucial for sustainable development and is closely connected to reducing poverty, enhancing education, and promoting gender equality. Lower-income populations, which depend on contaminated water sources and lack the means to implement protective measures, are disproportionately affected by arsenic-related diseases (Haque et al. 2018).

14.3.4 Global and Regional Attention

Recognizing the diverse consequences of As contamination, coordinated initiatives have been launched globally and regionally to tackle this issue. The joint monitoring program by WHO and UNICEF has underscored the importance of clean drinking water, stressing the need to improve water quality as part of the Sustainable Development Goals (SDGs). Multinational research endeavors, including the Arsenic Health Effects Research Program (AHERP) and the International Water Management Institute (IWMI), are dedicated to investigating and alleviating arsenic pollution.

14.3.5 Research and Development

Recent advances in the study of the origins, movement patterns, and biogeochemical cycles of arsenic have greatly enhanced our understanding of its environmental behavior and informed effective mitigation approaches. Progress in water purification methods, including the creation of more efficient and cost-effective filtration systems, has improved access to safe drinking water for affected populations (Hussam and Munir 2007). Moreover, eco-friendly farming techniques are being encouraged to limit arsenic absorption by crops and clean contaminated soils. The widespread and severe effects of arsenic on human health, the environment, and socioeconomic conditions have led to increased focus on its management within the framework of sustainable development. Successful management requires a comprehensive strategy that combines scientific investigations, technological advancements, community participation, and supportive policy structures.

14.4 Need for Arsenic Management

Effective arsenic management is essential to safeguard current and potential drinking water sources and reduce the harmful effects of arsenic exposure on humans and various natural resources. This significance is highlighted by the extensive nature of arsenic contamination, its serious health consequences, and its broad environmental and socioeconomic impacts. As illustrated in Fig. 14.4, arsenic management plays a vital role across multiple sectors in ensuring the health of various forms of life.

Fig. 14.4 Importance of arsenic management

14.4.1 *Ensuring Safe Drinking Water*

Arsenic contamination of groundwater poses a major threat to public health. Individuals relying on groundwater sources, such as wells and tube wells, especially in the South and Southeast Asian regions, have experienced adverse effects from arsenic contamination. Access to clean drinking water is crucial, as it is essential to sustain life. The World Health Organization (WHO) has established guidelines stating that arsenic levels exceeding 10 µg/L in water can be detrimental to any organism, leading to severe health consequences.

14.4.2 *Mitigating Health Impacts*

Prolonged exposure to arsenic in drinking water is associated with a wide range of health issues. The health consequences of exposure to arsenic can be severe and long lasting. For example, skin lesions and cancers caused by arsenic have high rates of disease and death, resulting in considerable human suffering and economic burdens (Naujokas et al. 2013). To address these health risks, it is crucial to implement effective management approaches, including providing safe alternative water sources, using arsenic removal methods, and consistently monitoring water quality (Hussam and Munir 2007).

14.4.3 Protecting the Environment

The environment is at considerable risk of arsenic pollution. Terrestrial and aquatic ecosystems can be affected by the accumulation of arsenic in soils and sediments (Rahman et al. 2019). In farming regions, crops, particularly rice, which is a primary food source in many affected areas, can absorb arsenic. This absorption can result in increased As concentrations in the food chain, thereby creating additional health hazards in humans. The well-being of aquatic organisms can also be compromised by arsenic, potentially leading to bioaccumulation and disruption of ecological balance (Sarkar et al. 2004). Comprehensive management strategies must not only tackle water contamination, but also prevent arsenic uptake by crops and protect overall environmental health.

14.4.4 Socioeconomic Implications

The widespread effects of As contamination on society and economy are significant. Communities affected by this issue often face financial difficulties due to decreased productivity caused by health problems and higher healthcare costs (Dasgupta et al. 2015). Access to clean drinking water is closely related to achieving sustainable development objectives such as reducing poverty, improving education, and promoting gender equality. In particular, women and children tend to bear the brunt of this problem as they are frequently tasked with collecting water and may need to travel longer distances to find safe sources, which in turn affects their educational and economic opportunities (Haque et al. 2018).

14.4.5 Global and Regional Efforts

The importance of effective management approaches for addressing As contamination has been underscored by both worldwide and regional initiatives. Joint monitoring programs and Sustainable Development Goals (SDGs) of organizations such as the WHO and UNICEF have stressed the need to improve water quality. Research and mitigation efforts to target arsenic contamination are the focus of international partnerships such as those established under the Arsenic Health Effects Research Program (AHERP) and the International Water Management Institute (IWMI).

14.4.6 Research and Development

Ongoing research and innovations are vital to improve methods for managing arsenic. Gaining knowledge of the geochemical properties of arsenic in various environments aids in forecasting and reducing contamination. Advancements in water purification techniques, particularly the creation of more effective and cost-effective filtration systems, are essential for supplying potable water to affected populations (Hussam and Munir 2007; Simon et al. 2024). Moreover, eco-friendly farming methods are being encouraged to limit arsenic absorption by plants and clean polluted soil. However, the importance of proper As management cannot be adequately emphasized. It is imperative to ensure safe drinking water, protect public health, conserve the environment, and promote socioeconomic growth. The resolution of this challenge requires a holistic strategy encompassing scientific studies, technological progress, community involvement, and supportive government policies.

14.5 Scientific and Regulatory Approaches for Arsenic Management

Previous studies have identified multiple approaches for managing As, which include therapeutic, regulatory, and community-oriented methods. These approaches play a vital role in tackling extensive contamination of groundwater and in reducing its harmful impacts worldwide.

14.5.1 Treatment and Assessment Technologies

Cutting-edge technological instruments play a crucial role in enhancing the precision and productivity of arsenic evaluation. Sophisticated equipment such as inductively coupled plasma mass spectrometry (ICP-MS) and X-ray Absorption Spectroscopy (XAS) enables the exact determination of As levels and forms in water specimens. Additionally, geographic information system (GIS) and remote sensing (RS) technologies contribute to the identification and surveillance of arsenic-prone areas, supporting focused intervention strategies. Various high-tech devices have been developed to extract As from potable water. These removal techniques can be broadly classified as physical, chemical, and biological approaches (Fig. 14.5). Arsenic can be removed from water using various methods.

a. Adsorption: This technique employs materials such as activated alumina, iron oxides, and modified activated carbon, which are strongly attracted to arsenic and can effectively lower its levels in water.
b. Ion Exchange: This process utilizes resins to exchange ions between the resin and water, which is particularly useful for water with a low to moderate arsenic

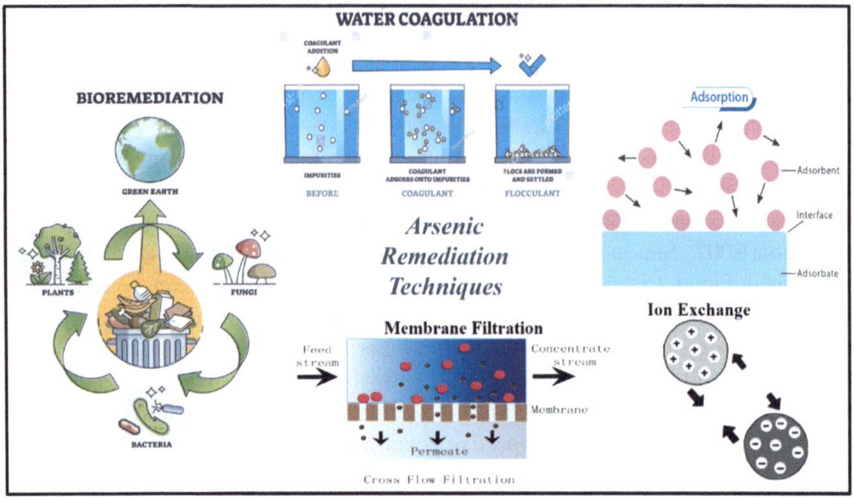

Fig. 14.5 Arsenic treatment techniques

content. It is often used in combination with other treatment methods (Clifford et al. 1998).

c. Membrane Filtration: Approaches such as reverse osmosis, nanofiltration, and ultrafiltration remove arsenic by passing water through a semipermeable membrane. Although highly effective, these techniques are expensive and require frequent maintenance (Waypa et al. 1997).

d. Coagulation and Filtration: This method involves the addition of chemicals, such as aluminium sulfate and ferric chloride, to water, forming flocs that absorb arsenic. The flocs were then removed through sedimentation and filtration. Large-scale water treatment facilities commonly use this technique.

e. Biological Methods: Bioremediation uses microorganisms to oxidize and precipitate As. Some bacteria can transform arsenite (As^{3+}) into the less harmful and more easily removable arsenate (As^{5+}). Although still experimental, this approach shows potential for sustainable removal of As (Majumder and Gupta 2015).

f. Sustainable Mitigation Technologies: To be widely adopted and successful in the long term, they must be affordable, culturally acceptable, and environmentally friendly. Promising options include iron-based adsorbents, phytoremediation, and solar oxidation and removal of arsenic (SORAS) in various settings (Pokhrel et al. 2005). Adapting these technologies to local conditions and involving the community in their implementation are essential to ensure their sustainability and effectiveness.

14.5.2 Regulatory Frameworks

Comprehensive policy frameworks and strong institutional backing are crucial for effective management of long-term arsenic contamination. Governments must implement and enforce drinking water, allocate funds for mitigation efforts, and support research on arsenic treatment technologies (Nordstrom 2002). Successful management requires cooperation among government entities, nongovernmental organizations, and international agencies. Regulatory actions should include establishing acceptable As limits in drinking water, assessing water quality, and ensuring compliance. The international guidelines established by the WHO set the allowable concentration of arsenic in water and serve as a benchmark for national standards, emphasizing the importance of consistent monitoring and risk evaluation. Numerous countries have adopted regulations based on WHO guidelines or developed their own standards. Effective As management requires regular testing of water sources to ensure adherence to regulatory standards. Governments and water authorities must implement robust monitoring programs and enforce regulations to protect public health.

14.5.3 Capacity Building for Arsenic Management

This research indicates that efficient dissemination of knowledge is crucial to improving capabilities in various contexts, including educational institutions and research partnerships. This plays a vital role in achieving sustainable growth. Building capacity serves as a key pathway to provide the general population with the skills and information to effectively handle arsenic contamination. This includes educational initiatives for regional water management officials, healthcare professionals, and community leaders to identify arsenic, assess risks, and implement reduction tactics. Enhancing technical proficiency among these groups improves the execution of effective As management methods, ranging from collecting and examining groundwater samples to creating and maintaining treatment systems. Community-based strategies are essential for tackling arsenic contamination, especially in rural areas and areas limited by resources. These approaches involve motivating and enabling local communities to take active steps to address As issues. Involving communities and increasing awareness of As exposure risks are essential elements of long-term management. Informational campaigns and interactive workshops help educate residents about As sources, their health effects, and safe water usage practices (Brammer and Ravenscroft 2009). Equipping communities with knowledge and resources allows them to take proactive measures, such as using arsenic-free water sources and implementing household filtration systems.

i. Raising Public Awareness: It is vital to inform communities about the dangers of As contamination and proper water usage practices. Informational initiatives, educational sessions, and skill-building programs can help communities in recognizing the importance of utilizing arsenic-free water sources and implementing suitable preventive measures.
ii. Public Engagement: Incorporating community members into the decision-making process ensures that arsenic management strategies are culturally appropriate and customized to meet local requirements. Collaborative approaches include community-driven monitoring, creation of localized water safety strategies, and formation of user groups to oversee water treatment systems (Ahmed et al. 2006).
iii. Tailored Approaches: Designing and implementing solutions that are appropriate for local conditions, such as household filtration units and community-scale water treatment facilities, can effectively address arsenic contamination. These solutions should be cost-effective, simple to maintain, and readily available to all members of the community (Hussam and Munir 2007).

14.6 Facilitating Knowledge Exchange Among Researchers, Policymakers, and Practitioners

Knowledge exchange between researchers, policymakers, and practitioners is vital for improving management approaches. Successful arsenic management requires an interdisciplinary strategy in which scientific findings inform policy choices and real-world experiences guide future research. This collaborative approach ensures that strategies are both scientifically accurate and practically viable. Soft skills of project managers, such as leadership and problem-solving abilities, are crucial for encouraging knowledge sharing within project settings. This is a key factor in the successful implementation of sustainable practices (Avença et al. 2024).

i. Connecting Research and Policy: A major management challenge lies in bridging the divide between scientific research and policy implementation. Although researchers often produce valuable data on arsenic contamination, its health effects, and mitigation techniques, this information may not always be readily accessible or comprehensible to policymakers. Similarly, policymakers may struggle to effectively communicate their requirements and limitations to the scientific community.
ii. Research Interpretation and Policy Summaries: Converting complex scientific discoveries into clear and concise policy summaries can help narrow this gap. These summaries emphasize key findings, public health implications, and suggested actions (Brownson et al. 2018). For example, the use of policy summaries in Bangladesh has played a crucial role in shaping national As mitigation programs (Ahmed et al. 2006).

iii. Cross-disciplinary Cooperation: Establishing platforms for cross-disciplinary cooperation can enhance communication and mutual understanding. For example, the Arsenic Research Group in Bangladesh unites researchers, government officials, and NGOs to discuss research outcomes and policy implications (Van Geen et al. 2003).

14.6.1 Integrating Local Knowledge and Practices

Practitioners, including local health workers, community leaders, and NGOs, provide valuable information on the local context, cultural practices, and practical challenges of arsenic mitigation. Integrating this knowledge with scientific research can lead to more effective and culturally appropriate interventions. Collaborative Research with Community Involvement (CRCI) engages local residents throughout the research journey by selecting research topics to gather data and share results. This method ensures that the study is rooted in the experiences of affected populations and that the results directly address their concerns. Recording and distributing examples and effective strategies from different areas can offer practical knowledge for other regions facing comparable issues. For example, the achievements of community-driven arsenic reduction initiatives in West Bengal, India, have been widely circulated to guide similar endeavors in other parts of South Asia (Chakraborti et al. 2011).

14.6.2 Enhancing Capacity Through Training and Workshops

Knowledge dissemination and stakeholder capacity building in arsenic management were significantly improved through training sessions and workshops. These events offer participants access to recent developments in arsenic research, remediation techniques, and regulatory frameworks. Specialized training programs that focus on cutting-edge arsenic detection and elimination methods can significantly improve the ability of water professionals to manage contaminated water resources. The World Health Organization (WHO) and the International Water Association (IWA) provide a variety of educational modules centered on water quality management (WHO 2017). Customized seminars designed for policy makers can enhance their understanding of the scientific principles underlying As contamination and the efficacy of various mitigation approaches. These gatherings also serve as forums for government officials to exchange experiences, discuss challenges, and participate in collaborative learning with peers.

14.6.3 Establishing Knowledge Networks and Platforms

Knowledge networks and platforms can facilitate continuous exchange of information and collaboration between researchers, policymakers, and practitioners. These networks include online platforms, regular conferences, and working groups.

i. Digital Knowledge Centers Web-based platforms, such as the Arsenic Knowledge and Action Network (AKAN), serve as repositories of information, discussion forums, and collaborative spaces for individuals involved in arsenic mitigation efforts.
ii. Symposiums and Conferences: Periodic gatherings focused on arsenic-related topics offer stakeholders the opportunity to present their studies, exchange experiences, and explore emerging challenges. For example, the International Conference on Arsenic in Groundwater played a crucial role in the dissemination of research results and encouraging cooperation.
iii. Committees and Task Forces: The formation of specialized groups consisting of members of academic institutions, governmental bodies, and community organizations helps ensure comprehensive and well-coordinated arsenic management strategies. These teams can focus on particular areas such as drafting policies, assessing technologies, and fostering community involvement.

14.6.4 Effectiveness of Knowledge in Arsenic Management

Several key elements influence the success of knowledge sharing and capacity building in arsenic (As) management. The suitability and efficacy of knowledge transfer, especially in cross-border collaborations, play a crucial role, as demonstrated in Sino-British educational initiatives (Li-Hua 2007). Furthermore, institutional stakeholders must understand the technical, geographical, and socioeconomic challenges to achieve sustainable As mitigation, as shown by their experiences in Bangladesh (Khan and Yang 2013). The effectiveness of capacity building interventions in public health practices, including arsenic management, depends on the selection of strategies that align with the priorities and context (Decorby-Watson et al. 2018).

Examining the socioeconomic aspects of arsenic management revealed intriguing contradictions. For example, arsenicosis is more prevalent among poor communities, suggesting a direct link between arsenic exposure and poverty (Nahar 2007). Furthermore, the idea of paying for water is not widely accepted in rural areas, which affects the sense of ownership and management of safe water sources (Khan and Yang 2013). The approach of environmental management to As contamination emphasizes the importance of focusing mitigation efforts on the most vulnerable populations (Singh and Taylor 2019).

The effectiveness of knowledge sharing and capacity building in arsenic management is complex and requires consideration of cultural, organizational, technical, and socioeconomic factors. It is crucial to adapt interventions to the specific challenges and needs of affected populations, ensuring appropriate and effective knowledge transfer (Li-Hua 2007) and a clear understanding of issues and mitigation measures among stakeholders (Khan and Yang 2013). Capacity building interventions must be carefully chosen to maximize impact, considering the socioeconomic context and the need for sustainable practices (Decorby-Watson et al. 2018). Addressing these factors is essential for successful mitigation of As contamination and public health protection. Information technology plays a vital role in disseminating knowledge about As threats and mitigation strategies (Rahoo et al. 2022).

14.7 Knowledge Gap

Although substantial efforts have been made in research, policy creation, and community involvement, there is still a shortage of holistic and unified strategies for managing As that adequately tackle the intricate social, economic, and environmental aspects involved. This gap in understanding presents considerable obstacles to effectively mitigating As contamination and its consequences. In organizational contexts, the importance of sharing knowledge in enhancing sustainable performance is emphasized, especially when affected by employee ambidexterity (Aamir et al. 2021).

14.7.1 Complexity of Arsenic Contamination

Arsenic contamination is a complex issue and is affected by numerous geological, hydrological, and anthropogenic elements. The behavior of As, including how it moves and spreads in groundwater, is intricate and varies by location, necessitating thorough site-specific research to comprehend and anticipate pollution trends. This intricacy is often overlooked in management approaches that frequently employ universal solutions instead of customized strategies based on particular local circumstances.

14.7.2 Social and Economic Dimensions

The management of As contamination often overlooks crucial social and economic factors. Communities affected by this issue, especially in economically disadvantaged areas, face substantial obstacles in obtaining clean water. These challenges include monetary limitations, insufficient infrastructure, and lack of technical expertise (Dasgupta et al. 2015). Furthermore, the far-reaching socioeconomic consequences of arsenic contamination, such as reduced productivity due to health issues and increased medical costs, necessitate comprehensive strategies that consider these wider implications (Rahman et al. 2019).

i. Prolonged exposure to As has significant health and economic consequences. Health impacts, including skin abnormalities, various forms of cancer, heart-related illnesses, and impaired development, place considerable strain on health-care systems and economic output (Naujokas et al. 2013). Effective management of these health issues requires extensive medical interventions and support networks, which are often insufficient in affected areas.
ii. Fairness and Accessibility: Providing equal opportunities to obtain clean water remains a significant obstacle. Many low-income families cannot afford the technologies used to mitigate water issues and there is often uneven allocation of resources and infrastructure. Although community-centered strategies show promise, they require expansion and support from strong institutional structures to ensure widespread and long-term access for everyone.

14.7.3 Environmental Considerations

i. The influence of arsenic on soil, crops, and ecosystems is a crucial but often neglected environmental concern. The accumulation in agricultural lands and crops, especially rice, which is a primary food source in many affected areas, can lead to increased exposure through the food supply. Consequently, effective As management must incorporate farming practices and consider broader environmental effects.
ii. Farming Methods: It is crucial to develop and encourage sustainable agricultural techniques that reduce arsenic uptake by crops. These approaches include creating arsenic-resistant plant varieties, enhancing irrigation methods, and implementing soil management strategies.
iii. Ecological well-being: As pollution can also negatively impact aquatic and land-based ecosystems, resulting in reduced biodiversity and ecological disruptions (Sarkar et al. 2004). Comprehensive management plans are required to address these environmental consequences and promote ecosystem health and adaptability.

14.7.4 Integrated and Multidisciplinary Approaches

To bridge this knowledge gap, it is crucial to employ comprehensive and interdisciplinary methods. This requires the fusion of scientific investigations with analyses of socioeconomic factors while also factoring environmental concerns into strategies for the management of arsenic. The creation and execution of successful solutions are highly dependent on joint efforts that unite scientists, government officials, field experts, and local populations (Silfia et al. 2023).

i. Cross-disciplinary Studies: Promoting research spanning the natural sciences, social sciences, and humanities can provide more comprehensive insight into arsenic pollution and its effects. This multifaceted approach may lead to the development of well-rounded management tactics that are scientifically accurate, socially acceptable, and economically feasible (Brown and Gilmartin 2011).
ii. Integrating Policies: It is essential to incorporate scientific discoveries into socioeconomic and environmental factors when formulating policies. This involves establishing practical and situation-specific regulatory guidelines, allocating sufficient funds and resources for implementation, and fostering cooperation across sectors (Chakraborti et al. 2011).
iii. Public Involvement: Including local communities in the decision-making process and incorporating their insights and experiences into management plans can improve the relevance and efficacy of interventions. Participatory approaches involving community members and initiatives that build local capacity are crucial to ensuring long-term sustainability and local ownership (Ahmed et al. 2006).

14.8 The Need for a Comprehensive Approach

To effectively manage arsenic contamination, a holistic and multifaceted approach is necessary that considers the social, economic, and environmental aspects of the problem. This claim is supported by the intricate nature of arsenic pollution and its widespread effects on communities, which demand diverse strategies for successful mitigation. Consequently, successful management approaches must be all-encompassing, incorporating technical, social, economic, and environmental elements, to produce lasting results.

i. **Technical Aspects**: Understanding the geochemical properties of arsenic, developing effective removal techniques, and implementing monitoring protocols are crucial technical considerations to address arsenic contamination in groundwater. Arsenic is primarily found in the Earth's crust in various forms, including arsenic sulfide, metal arsenate, and arsenide. Their mobility and distribution in groundwater are influenced by complex geochemical processes. The effectiveness of arsenic removal can be significantly affected by groundwater composition, particularly the ratio of iron, phosphorus, and arsenic concentrations. This ratio ([Fe] − 1.8[P])/[As], has been identified as a statistically significant control of As removal efficiency, regardless of the technology used. Artificial intelligence and machine learning models have been used to predict the geospatial distribution of this ratio in groundwater, potentially informing the selection of appropriate remediation strategies. This approach can help identify areas where sorption/co-precipitation technologies would be highly effective without iron amendment as well as regions where iron addition might be necessary for efficient arsenic removal.

ii. **Social Aspects**: Community-based participatory research (CBPR) and educational initiatives have been shown to be effective approaches to empower communities to take charge of mitigation efforts and develop sustainable solutions to local challenges. The CBPR emphasizes collaboration between researchers and community members throughout the research process, from identifying research questions to the implementation and dissemination of findings. This approach recognizes the unique strengths, knowledge, and expertise of community members to bring to the table, leading to more contextually appropriate and effective interventions. One of the key strengths of CBPR is its ability to address the specific needs and challenges of a community considering factors such as geography, history, culture, and social dynamics. This tailored approach can lead to more successful mitigation efforts, as demonstrated in a study in which a vulnerable community in Hawaii used PhotoVoice to identify and categorize its vulnerabilities and capabilities related to tsunami preparedness. This process resulted in the development of 16 feasible strategies to reduce the vulnerability of the community, highlighting the effectiveness of community participation in disaster preparedness. By involving community members at all stages of the research process, from problem identification to implementation of solutions, CBPR helps bridge the gap between research and practice. This approach not only leads to more effective interventions, but also empowers communities to continue addressing their challenges long after the initial research project has ended, ultimately contributing to improved health outcomes and reduced inequities (Ahmed et al. 2006).

iii. **Economic Aspects**: Financial limitations pose a significant obstacle to the implementation of arsenic mitigation technologies, particularly in economically disadvantaged areas. The cost-effectiveness of these technologies is a crucial factor in their adoption and sustainability. In Bihar, India, a cost-effectiveness analysis of three government-sponsored arsenic mitigation technologies revealed that a new hand pump (NHP) and new tube wells with stand-post (NTWSP) units were the most cost-effective interventions, while arsenic treatment units (ATU) were found to be the dominant but less cost-effective option. Interestingly, the choice of technology depends not only on cost, but also on socioeconomic factors. For example, NHP could be the most efficient intervention in areas with sparse and illiterate populations, while NTWSP could be more cost-effective in regions with higher literacy rates and arsenic awareness. This highlights the importance of considering local conditions when implementing arsenic mitigation strategies. Innovative approaches are being explored to address these limitations. One such approach is the micropayment-based systems-as-a-service model, known as "pay-a-penny-per-use" (PAPPU), which could catalyze the adoption of low-end, low-margin sensor analytic point solutions (SNAPS) for arsenic detection and mitigation in low-income markets. The development of locally available materials such as activated laterite, shows promise for large-scale implementation in community settings (Mondal and Garg 2017). These approaches demonstrate that while financial limitations are a significant obstacle, innovative solutions and careful consideration of local socio-economic conditions can help overcome these challenges in the implementation of arsenic mitigation technologies in economically disadvantaged areas (Hussam and Munir 2007).

iv. **Environmental Aspects**: Arsenic contamination has significant ecological consequences, affecting soil health, crop productivity, and human well-being. Soil arsenic pollution, resulting from both natural processes and human activities, such as mining and agriculture, disrupts the rhizosphere ecosystem and impacts overall soil functionality. This contamination not only decreases plant productivity but also degrades soil quality, affecting sustainable agricultural development. The use of arsenic-contaminated groundwater for irrigation exacerbates this problem, leading to Arsenic accumulation in soils and crops. This practice can lead to dangerous arsenic concentrations in the edible portions of crops, particularly rice, which is more susceptible to arsenic uptake than other crops. Interestingly, the long-term ecological consequences of arsenic pollution are often underestimated due to the absence of long-term monitoring data. A study of contaminated lakes in China revealed a significant increase in total arsenic levels, accompanied by pronounced ecosystem changes, including a substantial loss of crustacean zooplankton and an increase in metal-tolerant algae. Various strategies have been proposed to address these ecological consequences. Bioremediation approaches, including phytoremediation, microbial remediation, and microembolisation, offer eco-friendly and sustainable solutions to address As contamination and promoting soil health. Implementing these strategies, along with developing innovative technologies and sustainable mining practices, is crucial to mitigate As contamination and ensuring long-term ecological balance and agricultural sustainability. Multiple studies and successful examples support the idea that effective knowledge dissemination and skill development require a holistic and unified strategy.

Integrated approaches in Bangladesh that combine technical solutions with community involvement and policy support have substantially decreased arsenic exposure. These programs encompass multidisciplinary research, public education, and the deployment of cost-effective technologies (Ahmed et al. 2006; Hussam and Munir 2007). Research has shown that multipronged approaches to Arsenic mitigation are more successful than isolated efforts. For example, combining agricultural management techniques with water treatment methods has been shown to decrease Arsenic accumulation in crops and to enhance overall food safety (Norton et al. 2007, 2013). Global organizations such as the World Health Organization (WHO) and the United Nations Children's Fund (UNICEF) endorse comprehensive strategies that incorporate technical, social, economic, and environmental aspects in arsenic management (WHO 2017).

14.9 Conclusion

Research indicates that effective knowledge sharing is crucial for capacity building in various domains, including educational institutions and research partnerships, and is fundamental to sustainable development. The success of capacity-building initiatives depends heavily on the tailoring of knowledge-sharing practices to the specific requirements and contexts of the recipients. Furthermore, the role of knowledge sharing in improving sustainable performance, particularly when influenced by factors such as employee ambidexterity, emphasizes its importance in organizational environments. Despite these challenges, sharing tacit knowledge within companies, a key aspect of capacity building is essential for staff development and collective learning. International initiatives, such as STEP, showcase the importance of knowledge sharing in developing technical skills, including those needed for CCS technology, which can be applied to arsenic management. Project managers' interpersonal skills, such as leadership and problem-solving abilities, are vital for promoting knowledge sharing within project settings, which is critical for successfully implementing sustainable practices. The organizational culture significantly influences the facilitation of knowledge dissemination, a key element of knowledge management essential for gaining a competitive advantage. The influence of information technology on knowledge sharing among academic professionals further underscores the importance of technological infrastructure and competencies to allow effective knowledge exchange. Finally, the positive impact of knowledge management processes and knowledge sharing on teachers' literacy skills demonstrates the potential for these practices to improve educational outcomes, which can be utilized for sustainable arsenic management.

Acknowledgements The authors sincerely thank our prestigious institute Integral University and all involved in this study for their direct and indirect contribution.

References

Aamir, A., A. A. Nassani, A. Qadus, M. Haffar, and S. U. Jan, 2021. Impact of Knowledge Sharing on Sustainable Performance: Mediating Role of Employee's Ambidexterity. *Sustainability*, 13(22), 12788. https://doi.org/10.3390/su132212788

Ahmed, M. F., S. Ahuja, M. Alauddin, S. J. Hug, J. R. Lloyd, A. Pfaff, and A. van Geen, 2006. Ensuring safe drinking water in Bangladesh. *Science*, 314(5806), 1687–1688.

Avença, I., Domingues, L., and H. Carvalho, 2024. Do the Project Manager's Soft Skills Foster Knowledge Sharing? *Project Leadership and Society*, 5, 100139. https://doi.org/10.1016/j.plas.2024.100139

Brammer, H., and P. Ravenscroft, 2009. Arsenic in Groundwater: A Threat to Sustainable Agriculture in South and Southeast Asia. *Environment International*, 35(3), 647–654.

Brown, J. and Gilmartin, D. 2011. Groundwater, the Hidden Resource: The Historical, Hydro-Geological, and Socio-Economic Perspective. *Groundwater Governance: Global Frameworks of Action*. GEF-FAO Groundwater Governance Project.

Brownson, R. C., J. E. Fielding, and L. W. Green, 2018. Building Capacity for Evidence-Based Public Health: Reconciling the Pulls of Practice and Push of Research. *Annual Review of Public Health*, 39, 27–53.

Chakraborti, D., M. M. Rahman, B. Das, M. Murrill, S. Dey, S. C. Mukherjee and S. Roy (2011) Groundwater Arsenic Contamination and its Health Effects in India. *Hydrology Journal*, 19(4), 728–743.

Clifford, D. A., G. L. Ghurye, and A. R. Tripp, 1998. Arsenic in Drinking Water: An Update on the Treatment Technologies. *American Water Works Association.*

Dasgupta, S., M. Huq, M. Khaliquzzaman, K. Pandey, and D. Wheeler, 2015. Drinking Water Salinity and Arsenic in Rural Bangladesh: A Geospatial Analysis of Vulnerability. *Science of the Total Environment*, 488, 368–375.

Decorby-Watson, K., H. Manson, S. Abdi, K. Bergeron, G. Mensah, and B. Rempel, 2018. Effectiveness of Capacity Building Interventions Relevant to Public Health Practice: A Systematic Review. *BMC International Health and Human Rights*, 18(1). https://doi.org/10.1186/s12889-018-5591-6

Haque, S. S., A. M. Naser, M. Rahman, A. H. Milton, M. M., Hossain, M. H. Khan, and M. Rahman, 2018. Qualitative Assessment of Knowledge, Attitudes, and Practices Regarding Arsenic Mitigation in Rural Bangladesh. *Journal of Health, Population and Nutrition*, 37(1), 17.

Hussam, A., and A. K. M. Munir, 2007. A Simple and Effective Arsenic Filter Based on Composite Iron Matrix: Development and Deployment Studies for Groundwater of Bangladesh. *Journal of Environmental Science and Health, Part A*, 42(12), 1869–1878.

Khan, N. I., and H. Yang, 2013. Arsenic Mitigation in Bangladesh: An Analysis of Institutional Stakeholders' Opinions. *Science of the Total Environment*, 488–489, 493–504. https://doi.org/10.1016/j.scitotenv.2013.11.007

Li-Hua, R., 2007. Knowledge Transfer in International Educational Collaboration Programme. *Journal of Technology Management in China*, 2(1), 84–97. https://doi.org/10.1108/17468770710723640

Majumder, A., and S. K. Gupta, 2015. Arsenic Bioremediation Using an Arsenite-Oxidizing *Achromobacter* sp. Isolated from Arsenic-Contaminated Groundwater. *Journal of Environmental Science and Health, Part A*, 50(6), 568–576.

Mondal, M. K., & Garg, R. (2017). A comprehensive review on removal of arsenic using activated carbon prepared from easily available waste materials. Environmental Science and Pollution Research, 24, 13295–13306.

Nahar, N., 2007. Impacts of Arsenic Contamination in Groundwater: Case Study of Some Villages in Bangladesh. *Environment, Development and Sustainability*, 11(3), 571–588. https://doi.org/10.1007/s10668-007-9130-3

Naujokas, M. F., B. Anderson, H. Ahsan, H. V. Aposhian, J. H. Graziano, C. Thompson, and W. A. Suk, 2013. The Broad Scope of Health Effects from Chronic Arsenic Exposure: Update on a Worldwide Public Health Problem. *Environmental Health Perspectives*, 121(3), 295–302.

Nordstrom, D. K., 2002. Public Health—Worldwide Occurrences of Arsenic in Ground Water. *Science*, 296(5576), 2143–2145.

Norton, G. J., G. Duan, T. Dasgupta, M. R. Islam, M. Lei, Y. Zhu, and A. H. Price, 2013. Environmental and Genetic Control of Arsenic Accumulation and Speciation in Rice Grain: Comparing a Range of Common Cultivars Grown in Contaminated Sites Across Bangladesh, China, and India. *Environmental Science & Technology*, 47(22), 12243–12252.

Pokhrel, D., B. S. Bhandari, and T. Viraraghavan, 2005. Arsenic Contamination of Groundwater in the Terai Region of Nepal: An Overview of Health Concerns and Treatment Options. *Environment International*, 31(3), 312–316.

Rahman, M. M., N. Sohel, M. Yunus, M. E. Chowdhury, and A. Bhuiya, 2019. Socioeconomic Inequalities in Exposure to Arsenic-Contaminated Water: Results from the Health and Socioeconomic Survey in Matlab, Bangladesh. *Environmental Health*, 18(1), 1–10.

Rahoo, L. A., M. Memon, and S.-U.-H. Kandhroo, 2022. Impact of Information Technology on Knowledge Sharing Among Faculty Members of Public Sector Higher Education Institutions of Sindh, Pakistan. *Pakistan Journal of Social Research*, 4(3), 797–803. https://doi.org/10.52567/pjsr.v4i03.770

Sarkar, A., B. Paul, and S. Choudhury, 2004. Arsenic Contamination in Groundwater of North Eastern India. *Journal of Environmental Science and Engineering*, 46(2), 162–174.

Silfia, A., R. Purwono, and N. Herachwati, 2023. Knowledge Sharing: Antecedent and Consequences Literature Review. *RSF Conference Series: Business, Management and Social Sciences*, 3(3), 319–330. https://doi.org/10.31098/bmss.v3i3.678

Simon, M., Joshi, H., Yadav, A. K., & Hopke, P. K. (2024). Simultaneous nitrogen and phosphorus removal from polluted water using a novel bacterial consortium (BACON-38): Promising application for wastewater remediation. Journal of Water Process Engineering, 66, 106044. https://doi.org/10.1016/j.jwpe.2024.106044

Singh, S. K., and R. W. Taylor, 2019. Assessing and Mapping Human Health Risks Due to Arsenic and Socioeconomic Correlates for Proactive Arsenic Mitigation. In *Springer* (pp. 231–256). https://doi.org/10.1007/978-3-030-21258-2_10

Sobandi, A., M. A. Ramdhany, R. Rasto, and E. Suryadi, 2021. Knowledge Management Process, Knowledge Sharing, and Teacher Literacy Skills at Vocational High Schools. *Jurnal Cakrawala Pendidikan*, 40(3), 738–749. https://doi.org/10.21831/cp.v40i3.42489

Van Geen, A., Y. Zheng, Z. Cheng, Y. He, R. Dhar, J. M. Garnier, and K. M. Ahmed, 2003. Impact of a Large-Scale Switch from Groundwater to Surface Water on Arsenic Epidemiology in Bangladesh. *Bulletin of the World Health Organization*, 81(12), 788–793.

Waypa, J. J., M. Elimelech, and J. G. Hering, 1997. Arsenic Removal by RO and NF Membranes. *Journal - American Water Works Association*, 89(10), 102–114.

World Health Organization (WHO), 2017. Arsenic in Drinking Water. https://www.who.int/news-room/fact-sheets/detail/arsenic

The manufacturer's authorised representative in the EU is Springer Nature Customer Service Centre GmbH, Europaplatz 3, 69115 Heidelberg, Germany. If you have any concerns regarding our products, please contact ProductSafety@springernature.com

Printed and bound by CPI Group (UK) Ltd, Croydon, CR0 4YY
30/09/2026
02225564-0005